Bird Problems in Agriculture

Bird Problems in Agriculture

The Proceedings of a Conference
"Understanding Agricultural Bird Problems"
held at Royal Holloway College, University of London
April 4th–5th 1979

Edited by
E. N. WRIGHT, B.Sc., F.I.Biol.
assisted by
I. R. INGLIS, Ph.D. and C. J. FEARE, Ph.D.

BCPC Publications
144–150 London Road, Croydon CR0 2TD

144–150 London Road
Croydon

Bird problems in agriculture
Bibl.-Index
ISBN 0 901436 48 8
1. Title 2. British Crop Protection Council
3. Wright, E. N. 4. Inglis, I. R. 5. Feare, C. J.

Printed in Great Britain by
LAMPORT GILBERT PRINTERS LTD
Wantage Road, Reading, Berkshire

Contents

Preface

Agricultural bird problems are not a new phenomenon, although their importance to individual farmers is undoubtedly heightened by the economic pressures of modern life. Historically there are many references to birds attacking man's crops and it seems certain the problem is as old as agriculture itself. Paintings in the tombs of the Pharaohs are reputed to depict ancient Egyptians scaring birds and the decoration on some Greek and Roman pottery unequivocally shows birds eating cherries.

In the Middle Ages explicit references to the damage done by birds to agricultural and horticultural crops are to be found in early enactments on pests. King James I of Scotland introduced an Act for the destruction of rooks in 1424 and in England a century later, in 1533, an Act was passed ordering Parishes to set nets to capture choughs (jackdaws), crows and rooks and imposing heavy fines for non-compliance. Presumably these measures failed to produce the desired effect for in 1566 another Act introduced bounties, of varying amounts, to be paid by churchwardens, on a list of birds and mammals described as "vermin". Books on agriculture and horticulture published around this time also refer extensively to pests and although there is sometimes confusion over the names of species (Linnaeus did not bring order out of chaos until the mid-eighteenth century) the authors often show a remarkable insight into the nature and complexity of the problems and their possible solution. Gervase Markham was a prolific writer on matters concerning farming and the countryside and in his book *Markham's Farewel to Husbandry*, published in 1668, he had the following to say about bird pests and ways of preventing damage to crops.

Book II. *kind of Grain.* **73**

Crows, Pidgeons, and Birds.

To begin then with the firſt enemies of corn or grain, after it is thrown into the earth, there is none more noyſome then *Crows*, and *Choughes*, and other ſmaller birds, which flocking after the ſeeds man will in a manner devour and gather up the grain as faſt as it is ſown : for as according to the old ſaying, That *many hands make light work*, ſo many of their mouths (being creatures that ever flye in flocks together) and their much nimbleneſs in devouring, ſoon rob the earth of her ſtore, and deprive the labouring Husbandman of very much profit, and the Grain which the creatures do moſt conſume, is all manner of white Corn, as Wheat of all kinds, Rye and Oats, as alſo Hemp-ſeed, Lin-ſeed, Rape-ſeed, and ſuch like : Neither are they only offenſive during this time of ſowing, but after it is ſown and covered, digging it with their ſtrong bills out of the earth, and ſo making the waſte greater and greater.

The Cure.

The prevention or cure of this evill, is divers, as the affecti-ons of people, and cuſtom of Countries do inſtruct them : for

ſome

ſome (eſpecially the *French-men*) uſe when they ſow theſe grains and ſeeds, firſt to ſprinkle it with the dregs or lees of their bittereſt oyles, which when theſe devouring fowls do taſte, they refuſe to do any further hurt: Others uſe to ſow Pigeons dung or Lime with their ſeed, which ſticking unto the grain, the unſavorineſs thereof will make the fowl caſt up the grain again, and leave to do further hurt. But foraſmuch as theſe medicines cannot ever be had, nor are ever wholſome for every ground, the onely beſt and ſafeſt means to prevent this evill, is; to have ever ſome young boy, with bow and arrows to follow the ſeed-man and Harrows, making a great noiſe and acclamation, and ſhooting his arrows where he ſhall ſee theſe devourers leight, not ceaſing, but chaſing them from the Land, and not ſuffering them at any time to leight upon the ſame, and theſe ſervants are called Field-keepers, or Crow-keepers, being of no leſs uſe and profit (for the time) then any other ſervants whatſoever. Nor is it ſufficient to have the Field-keepers for the bare time of ſeed only, whileſt the grain is in ſowing, but he ſhall alſo maintain them untill ſuch time as you ſee the grain appear above the earth, which for Wheat or Rye, becauſe they are winter ſeeds, and ſo longer in ſprouting, will ask a full month; for all other ſeeds which are ſown in the Spring, or Summer, a fortnight is full ſufficient: And this Field-keeper ſhall not fail to be in the field an hour before Sun in the morning, and ſo continue till half an hour after Sun-ſet in the evening, for at the riſing and ſetting of the Sun is ever done the greateſt miſchief, for then are all creatures moſt eager and hungry; and though the indurance may promiſe much pain and trouble, yet queſtionleſs the labour to any free ſpirit, is both eaſie, and pleaſant.

Additions. Alſo if your Field-kepeer, inſtead of his bow and arrows, do uſe to ſhoot off a musket, or Harquebuſh, the report thereof will appear more terrible to theſe enemies of Corn, and the profit thereof will be a great deal more: for a ſhot or two of powder will ſave more Corn, than a weeks whooping and ſhouting, onely you muſt obſerve, that your Field-keeper uſe no bullet or hail-ſhot, for ſo he may turn ſcaring to killing. Now touching the deſtruction which theſe creatures make of Corn after it is is ſtackt up by tearing off the thatch, and digging holes and pits therein; to prevent that, you ſhall cauſe the thatcher to ſcatter upon the thatch, great ſtore of aſhes of any kind, or elſe Lime, that as the Pidgeons or Crows tear up the ſtraw, the Lime or aſhes will ſparkle into their eyes and nares, which they will not indure; as for thoſe parts of the ſtack which cannot be thatcht, as the ſides and ends; upon them you ſhall prick divers ſcare-crows, as dead Crows, or dead Pidgeons, or any other rags, as the ſhape of a man, made either of thumb ropes of hay or ſtraw, or elſe ſome old caſt away apparrel, ſtopt with ſtraw, and ſo fixed on the ſtack, alſo in this caſe you may uſe Clap mils, or ſuch like toyes, which make a great noiſe. But to conclude, the beſt prevention for theſe creatures (if you want ability to maintain a field-keeper) is to take long lines of pack-thread, and in them to knit divers feathers of divers colours, eſpecially white ones, and with little ſtakes ſo faſten them over the Corn, that with every breath of wind the feathers may dance and turn about, and the nearer that theſe blinks or ſcares come to the ground (when the corn is new ſown) ſo much the better it is, leaſt the fowl, finding a way to creep under them, begin not to reſpect them; ſo that a hand or two from the ground is ſufficient, Provided that the feathers and ſcares have liberty to play and move.

But if it be to ſave Corn in ripening, that is to ſay, a little before it be reapt, when the ears begin to harden, or when it lyeth in ſingle ſheaf upon the land, for then fowl and birds do as great miſchief, as at any other ſeaſon, it ſhall then be fit that you raiſe theſe lines or ſcares upon higher ſtakes, ſo as they may play as much above the ears of the corn, as before they did above the earth; and amongſt theſe ſcares thus made upon lines in ſundry parts of the field, you ſhall upon other ſtakes place many other bigger ſcares, as dead Crows, Pies, Gleads, Pigeons, or ſuch like, as alſo the proportion of a man formerly ſhewed you, or any rags of cloath being black, fowl, and ugly, like bakers malkins; and than this, there is no ſafer way for the defence of grain, or corn, from theſe birds, and ſuch like.

I have quoted Gervase Markham at length because of the striking relevance of his remarks to topics discussed at this symposium. It is impossible to read this account of the situation in the mid-seventeenth century without becoming acutely aware of the lack of progress in bird damage control over the past 300 years. The traditional methods that are often still used are relatively inefficient and labour intensive and, in the modern world, are neither practicable nor economic. Farmers everywhere are crying out for more cost-effective ways of protecting their crops against birds but the biological, environmental, social and legal obstacles are formidable and progress is slow. The primary reason for holding this symposium, therefore, was to stimulate interest in seeking new approaches to the control of bird damage.

The British Crop Protection Council was founded in 1968 by the amalgamation of the British Weed Control Council with the British Insecticide and Fungicide Council. The members of the Council, although drawn from different organizations, have a common objective in the promotion and encouragement of the science and practice of pest, disease and weed control. In 1975, in response to the growing concern of the farming industry over bird damage, the BCPC departed from tradition by including a session on birds in the programme for its biennial crop protection conference. This proved so successful that a session on vertebrate pests was again included in the 1977 conference. Despite the interest shown in these sessions it was apparent to the organizers that the nature of the subject demanded a more substantial hearing than could be catered for in the format of those meetings; Council therefore decided to hold a two-day symposium on the specialist topic of bird damage.

From the outset it was intended that the symposium Understanding Agricultural Bird Problems should form a springboard for future action in the fields of research and development. Hence a structured approach was favoured and attendance was limited to one hundred invited delegates chosen to represent a cross-section of people with relevant interests in birds and farming. The meeting was arranged around four sessions, each having a theme, and I am indebted to my co-editors, Chris Feare and Ian Inglis, for organizing sessions on The Starling and Bird Scaring respectively. I should also like to thank all who contributed, whether as chairmen, speakers or participants in the discussions; I hope they will find satisfaction in the publication of these collected papers. Finally, since the success of any conference depends ultimately on the extent to which people talk to each other, the venue was selected with some care in an attempt to provide an atmosphere in which discussion could flourish. The Royal Holloway College proved a happy choice, combining relative seclusion with ease of access; good facilities with willing and efficient service. I should like the domestic staff of the College to know that their efforts were much appreciated.

E. N. Wright

GENERAL CONSIDERATIONS

Introduction

E. N. Wright
Ministry of Agriculture, Fisheries and Food
Worplesdon Laboratory

Ladies and gentlemen. It is with very great pleasure that I welcome you to this symposium on Understanding Agricultural Bird Problems, organised on behalf of the British Crop Protection Council. Our aim in arranging a closed meeting of this kind had two objectives. Firstly, to bring together people – farmers, academics, technologists, advisers, conservationists and administrators – chosen for their ability to contribute a cross-section of knowledge relevant to the central issue of how to prevent birds from damaging crops. Secondly, by restricting the number of delegates it was hoped to promote an atmosphere in which discussion could flourish. I believe these organisational objectives have been achieved; the success of the symposium is in our own hands and will be judged by the quality of our deliberations.

Whilst our primary concern is with British agriculture we must remember that similar bird problems occur in other countries and the potential value of this meeting is enhanced by the presence of delegates from overseas. All too often we allow national considerations of political, social or economic origin to obscure the international nature of biological problems; this is especially true of birds, whose migrations mock the territorial behaviour and aspirations of man.

In our first session we shall be concerned with a number of considerations that form a background to any attempt to manipulate bird populations and we start on a conservation theme. With 80% of land use in Britain devoted to agriculture most of our wildlife must, perforce, live and feed on farmland; therefore, what size wildlife populations can we afford? Modern farming is the antithesis of the image projected by the old adage about sowing corn – one for the rook; one for the crow; one to rot and one to grow. Nowadays the farmer sows "one to grow"; competition within the industry allows little margin for natural wastage. One way out of the dilemma, at least for Western society, is to compensate farmers for damage caused by wildlife and we shall hear of Canadian attempts to operate such a scheme in respect of wildfowl damage to cereals. A starving society would not be free to make such a choice but, even if we were mindless of the need for conservation, could we really control populations of bird pests? Jim Flegg considers some of the biological factors affecting control strategy. In the final session this morning Tony Colling discusses man-made constraints on bird control in the context of the Council of Europe Directive on the Conservation of Wild Birds.

How many wild birds should farmland support?

N. W. Moore
Nature Conservancy Council
London

Summary

At present, farmlands in Britain probably support between 50 and 100 million pairs of breeding birds belonging to about 130 species. It is desirable on many grounds that farmland should continue to support viable populations of these species. Most (62%) species have relatively small total populations (<100 000 pairs) and could be severely threatened if their habitats suffered yet further large-scale reductions or they were damaged by pesticides etc. On the other hand most "pest species" occur in very large numbers and could be reduced without detriment. However, control measures must be specific if they are not to damage more vulnerable species which feed in the same area. The most difficult problem is provided by winter flocks of relatively rare species which cause local damage.

Introduction

Conservation involves the deliberate control of populations of organisms. Under the circumstances which exist today most species require positive measures to ensure their survival; the main problem is how to provide an adequate amount of suitable habitat for them. However, at certain times and in certain places the populations of some species, about 7% of the total, need to be reduced because they cause damage to agriculture or to other species of wildlife. Most of these pest species have populations of one million or more pairs and their control is the subject of this symposium. The aim of this introductory paper is to put the control problem in the wider context of conservation. This may be particularly worth doing because administrative arrangements put different aspects of bird control under different environmental and agricultural departments and this obscures the essential unity of the problem. The question asked in the title of this paper is fundamental to all who seek to control bird populations.

The short answer

The question put to me – "How many wild birds should farmland support?" – could refer to species or to numbers: I shall assume it means both. I also assume that farmland is taken to mean the complex of habitats whose produce provides a livelihood to farmers, that is about 80% of the land surface of Great Britain. Farmland excludes urban areas, large woodlands and plantations, large lakes and reservoirs, coastal areas and uplands which are not grazed by sheep or cattle. Thus seabirds, some duck and some waders are excluded.

I suggest that British farmland should support about 130 species of birds and that the numbers should be of the order of 50 million pairs. The arguments in support of these figures are outlined below.

Number of species

The number of species required depends upon the aims of biological conservation. Conservation is supported by several motivations – scientific, aesthetic, economic and educational. The common denominator is the need to maintain the living natural resource which has been received from the past and which, if man wills it, can continue to evolve in the future. The resource can be looked at in terms of ecosystems, species or genetic information, but the essence of conservation is the maintenance of diversity.

This requirement is obvious from the scientific and aesthetic viewpoints; it may be less so from the agricultural. Yet only a minute proportion of wild species have been domesticated by man – for example, only two British birds, the greylag goose (*Anser anser*) and the mallard (*Anas platyrhynchos*) – and it would be wrong to deny future generations the possibilities of further domestications: conservation is essentially about the future. We have to conclude that in general terms it is desirable to retain as many species as possible.

About 230 species of wild birds breed in Britain: of these about 130 (57%) are wholly or largely supported by farmland and it is desirable to retain these species for the reasons give above. Doubtless some new breeding species will colonise Britain as a result of small climatic changes, as the redwing (*Turdus iliacus*) has done in recent years, and others will become extint here – perhaps the redbacked shrike (*Lanius excubitor*)? A few, like the little ringed plover (*Charadrius dubius*), may be able to colonise new man-made habitats and a very few may undergo a genetic change which allows them to colonise areas not formerly available to them, e.g., the collared dove (*Streptopelia decaocto*); but generally speaking the number of available species is unlikely to change very much in the short- and middle-term future. The extent to which these 130 species can survive on farmland will increasingly depend upon how farmers farm, and that will depend to a large extent on fiscal and economic contraints determined by Government.

Number of individuals

The existing farmland species can only be conserved if their populations are large enough to be viable. Despite a considerable development of island biogeographical theory, which is relevant to habitat islands as well as geographical ones, and despite considerable empirical knowledge, it is rarely possible to state, even approximately, the minimal viable size of a population of a given species. Some species tick over at remarkably low levels, for example the kite (*Milvus milvus*) in Wales, and the Dartford warbler (*Sylvia undata*) in southern England, but others may reach a point of no return while still apparently numerous, e.g., the American passenger pigeon (*Ectopistes migratorius*). In the circumstances the conservation biologist must be conservative and recommend that a population size which has proved its viability in the past should not be seriously reduced. Therefore the less common species, i.e., those whose population is 100 000 pairs or less, should not be allowed to become much rarer, while for the commoner ones, those exceeding 100 000 pairs, considerable reductions may be acceptable in instances where they can be proved to be serious pests. The order of magnitude figure of 50 million pairs mentioned above is derived from making these assumptions (see Table 1).

Table 1

Population sizes of farmland birds in Great Britain

Population category	Number of species in category	Estimate of numbers of pairs in category
>1 000 000	17	56 000 000
100 000–1 000 000	32	10 000 000
<100 000	81	2 000 000
Totals	130	68 000 000

Note on calculation of population estimate
*An assessment was made of the Irish population and this was subtracted from the mean for the British Isles, given by Sharrock (1976), in order to obtain the Great Britain estimate. It should be noted that about 30 species are found on these types of land as well as on agricultural land, e.g. oystercatcher (*Haematopus ostralegus*), and so the agricultural values for these species are inflated, but since the total number in this mixed habitat group is less than 1 million pairs this has little general significance.*

Of course the figure of 68 000 000 for the existing number of pairs of farmland birds is a very rough estimate, but it does have value in showing:

(a) the order of magnitude of the total population of birds which farmland supports and could be expected to support in the future;

(b) the abundance of some species, and hence the acceptability on conservation grounds of control of those species, which are harmful to agriculture: the estimated British population of the starling (*Sturnus vulgaris*) is c. 4 000 000 pairs, of the house sparrow (*Passer domesticus*) c. 4 500 000 pairs, of the skylark (*Alauda arvensis*) c. 2 500 000 pairs and of the woodpigeon (*Columba palumbus*) c. 3 000 000;

(c) the large number of species whose total populations in Britain may be precarious because of their small size. This emphasises the need to ensure that measures aimed at reducing the problem species do not have significant effects upon these rarer species.

The farmland environment in relation to desirable stocks of wild birds

The overall aim must be to maintain viable stocks of all species, and at the same time to ensure that none of the very abundant species becomes seriously damaging to agriculture. To achieve effective conservation the origin of the British avifauna has to be considered. Originally most of the country was covered by forest and there were extensive marshes; open ground, which consisted of a few mountain tops, and coastal habitats, such as sand dunes, made up a very small amount of the total. Thus, numerically, both as individuals and species, birds of woodland and marsh predominated.

Since the Neolithic period most of the forest land and nearly all the marshes have been reclaimed for agriculture, and thus open-ground habitats have increased from perhaps less than 5% to over 80%. Broadleaved forest, the climax vegetation, today covers less than 5% of the land, and has been profoundly altered by management. We have no evidence about the woodland species which became extinct when the forests were cleared, but obviously the numbers of woodland birds must have declined enormously. The decline of marshland led to the extinction in Britain of several species, e.g., the crane (*Grus grus*), although some of them, e.g., the bittern (*Botauras stellaris*) and black-tailed godwit (*Limosa limosa*), have since recolonised the country following the introduction of protection laws. Relatively few species have been able to exploit the opening up of the land, but those that have have mostly become very abundant, e.g., the lapwing (*Vanellus vanellus*), skylark (*Alauda arvensis*) and partridge (*Perdix perdix*). Despite immense changes in land use, most British land birds still depend on broadleaved woodland, although hedges and conifer plantations provide a habitat, albeit one that is frequently sub-optimal, for some species.

In the period since the second world war further changes in forestry and agricultural practice have had profound effects on British wildlife. The most important from the bird point of view have been:

(a) the conversion of broadleaved woodland to arable;
(b) the conversion of broadleaved woodland to conifers;
(c) the reduction of the total acreage (mileage) of hedgerows;
(d) the reclamation of heathland and scrub;
(e) improved drainage of large areas of low lying land (e.g. Romney Marsh);
(f) the greatly increased use of pesticides.

The loss of habitat varies from area to area. In some Scottish counties the loss of woodland has been over 50%, in Cambridgeshire it is 17% (for a brief summary on habitat changes see Nature Conservancy Council 1977). The use of persistent organochlorine insecticides caused substantial reductions in the populations of several birds of prey (for summaries on individual species see Sharrock (1976). Following restrictions on the use of organochlorine insecticides, most of the affected raptor populations are now recovering but, with the exception of the partridge (see Potts 1970), the effects of pesticides on most other birds have not been studied.

If farmland is to retain its bird life in anything approaching its present variety and abundance, the following measures are required:

1. Fiscal and economic measures to halt the destruction of broadleaved woodland.
2. Conservation measures, backed by adequate finance, to retain and manage the more important wetlands and heathlands primarily for wildlife.
3. Continued surveillance of pesticides to ensure that no compound causes serious damage to bird populations.

In the short term at least, polarisation of land use seems inevitable: technological advances in agriculture will make the crop lands less and less suitable for most species of wild birds, and the losses will have to be made good by retaining and managing marginal habitats so that they can support more wildlife. As polarisation increases, and as areas of marginal habitat decline, the few bird species which can adapt to the habitats produced by new agricultural technologies, are likely to become serious pests, and measures will have to be taken to control them. Many species that depend on marginal habitats feed on croplands, e.g., the chaffinch (*Fringilla coelebs*) and will be potentially at risk from control measures directed against pest species. It is therefore essential that the control measures are selective in their operation.

Winter visitors

So far we have only considered breeding birds. Owing to its mild climate, Britain receives very large numbers of winter migrants. Most do no damage to agriculture, but on occasions wild geese may harm young corn and grass. Wild geese are relatively rare animals, but because they form large winter flocks they can have serious local effects. Extensive reductions in numbers are undesirable from the conservation point of view, yet individual farmers require practical help in overcoming damage to crops. The problem is discussed later in this symposium. When it is not possible to decoy geese away from valuable crops, adequate compensation for damage may be the only solution to this difficult problem.

References

NATURE CONSERVANCY COUNCIL (1977) Nature Conservation and Agriculture. London; Nature Conservancy Council, 40pp.

POTTS, G. R. (1970) Recent changes in the farmland fauna with special reference to the decline of the Grey Partridge. *Bird Study* 17, 145–166.

SHARROCK, J. T. R. (1976) The atlas of breeding birds in Britain and Ireland. Aylesbury; Hazell, Watson and Viney Ltd. 477pp. ISBN 0 903793 01 6

Biological factors affecting control strategy

J. J. M. Flegg
East Malling Research Station
Maidstone

Summary

Bird species associated with crop damage are usually numerous and widespread, and often only a small proportion of the population causes damage. Most bird pests have a flexible diet, in which crop plants may feature only irregularly or incidentally, and, in common with other species, they have a high reproductive capacity. Many migrate over considerable distances. The speed and time of these migrations depends largely on climatic factors but sometimes on other considerations such as population density. Even in those species normally classed as sedentary, there may be enough local movement to allow infilling after a proportion of the population has been eliminated. A lack of knowledge of these, and other factors, impedes the design of control strategies.

Overall, before effective damage-control strategies can be designed, much more detailed information is needed. The general ecology of pest species in proximity to crops, the role of the various items in their diet and their behaviour all require investigation. Studies to obtain such information may often be specific to one pest/crop/locality combination. For many avian pest situations, more forethought, based on ecological considerations, in positioning crops or in land-management techniques may effectively reduce the cost of damage. It would also reduce the difficulties inherent in devising effective strategies for crop protection by the use of deterrents, population control measures, or physical means, against adversaries which, by their evolution and behaviour are well-equipped to exploit our crops.

Introduction

In devising strategies to defend crops from birds, underlying biological factors, possibly better described as hazards, are too often overlooked. These factors and their adjuncts are legion, so this review must necessarily be selective.

The term "control strategy" is broadly used here and embraces tactics such as the attempted massacre of the offending species and the use of various means of deterring pests from attacking crops. It also includes taking note of the environmental needs of problem and potential problem species, and endeavouring to deny them these resources by planning ahead. On the assumption that pest organisms are, in general, pre-adapted to exploit some part of the man-made agricultural environment, it follows that, in an era when land-management is changing rapidly, we should expect new cropping techniques to be vulnerable to the flexibility of old-established pests or to open the door to new ones.

The biological factors that need to be taken into account when devising effective control techniques can be considered under two broad headings:

(i) those associated with the nature of the habitat and its climatic environment, and
(ii) those associated with the nature and biology of the pest itself.

Naturally, there is often no clear line dividing the two.

Habitat

An excellent example of the sort of problem that can arise from habitat manipulation – an appropriate introductory "cautionary tale" – comes from studies of the problems caused by birds at airfields. At Vancouver Airport, Canada, the short-mown grass attracted flocks of gulls and other birds, seeking a safe roosting or feeding area. These flocks became a hazard to aircraft and in an attempt to make the airport less attractive to birds the grass was allowed to grow tall. This was effective, but in the resultant long grass, small rodent populations expanded rapidly, attracting predatory birds to the extent that over 500 (mostly short-eared owls (*Asio flammeus*)) were trapped by the airport management over a period of three years (Lewis, 1967).

In the field of horticulture, recent advances have brought about a great improvement in the "cleanliness" of orchards: herbicides are used to maintain vegetation-free strips of soil adjacent to the trunks, and efficient mowing, together with the use of herbicides, has greatly reduced the intrusion of most broad-leaved plants into the grassed alleys. Furthermore the grass itself rarely sets seed. In soft-fruit crops also the increased use of herbicides maintains relative freedom from weeds. With this steady reduction in "natural" food resources, should we be surprised at a concurrent increase in seed removal from strawberry fruits by linnets (*Acanthis cannabina*)? In some areas, this problem is entering a new phase; the introduction of polythene tunnel cloches has produced even earlier crops, which provide a readily available food source at a time of natural scarcity. Linnets have discovered the crop under the polythene, and are enjoying the food and the insulation that the tunnels provide from both the elements and any deterrent devices that the grower may employ.

The period since the war has also seen the removal of a considerable mileage of hedgerow and with it a broad-spectrum of food plants. In orchards, the replacement of hedges with windbreaks of alder (*Alnus* spp.) may also bring problems in its train. Redpolls (*Acanthis flavirostris*), which are subject to erratic population fluctuations, are often part of the orchard avifauna and can cause damage to opening blossom and fruitlets (Stenhouse 1962). The seeds of alder are a major natural food resource for the redpoll: therefore it is pertinent to ask whether alder windbreaks are playing a part in the current upsurge of redpoll numbers (Sharrock, 1976)? Are we creating a problem for the future by spreading this food resource, thereby attracting redpolls to orchard localities, where, when the alder seed runs out, buds and fruitlets are readily available?

This opens a new field for discussion: the location of a sensitive crop in the usually diverse farmland environmental mosaic. Bullfinches (*Pyrrhula pyrrhula*) are considered to be forest birds ancestrally (Newton 1972) but have proved successful in Britain as birds of broken woodland, penetrating into adjacent farmland. The bullfinch is on record as a pest of fruit crops since the sixteenth century, damaging buds of gooseberry, black currant, plum, pear and some apples; the pattern of damage is also well established, initial attacks being on the plantation rows adjacent to woodland cover. At the present time pears, especially the varieties Williams' and Conference, are among the major targets, but despite this predictability, a surprising number of pear orchards are still located in close proximity to woodland and are damaged accordingly (Fig. 1). On ecological grounds, there would seem to be a good case for attempting to minimise damage by planting pear orchards amidst blocks of non-favoured apples and as distant as possible from any woods which may harbour bullfinches.

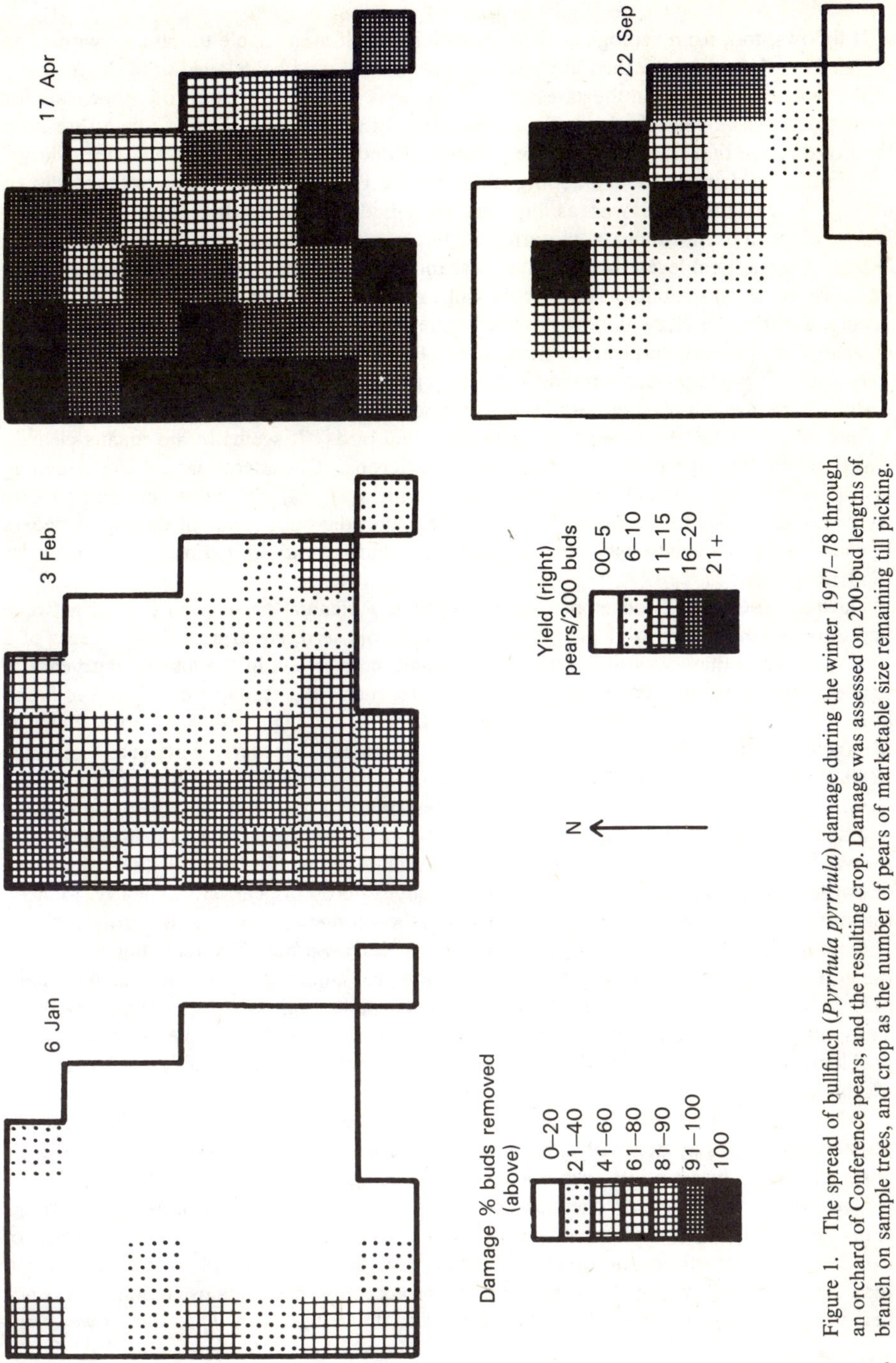

Figure 1. The spread of bullfinch (*Pyrrhula pyrrhula*) damage during the winter 1977–78 through an orchard of Conference pears, and the resulting crop. Damage was assessed on 200-bud lengths of branch on sample trees, and crop as the number of pears of marketable size remaining till picking.

It follows, too, from ecological considerations, that if areas liable to damage within the orchard can be predicted, then in these areas pruning should be delayed until the measure of the season's damage can be assessed. This at least allows the grower to compensate for damage, as far as possible, by judicious pruning. In pears, there seems to be a threshold level of damage that trees can tolerate (Smith & Kendall 1975). This has also been shown by stimulating damage by disbudding (Summers & Pollock 1978). In the orchard shown in Fig. 1, this threshold was as high as 80% bud loss in 1978 (Flegg & Matthews 1979). Since it is normally a minority of trees in an orchard that suffer bud loss in excess of the critical level, the remainder escaping relatively unscathed, if the damage can be more evenly distrubuted fewer trees will exceed the threshold at which loss of crop occurs. Clearly, therefore, it is worthwhile attempting to do this by concentrating deterrent measures in the peripheral areas most likely to be damaged.

Is our land management "forcing" wild bird populations increasingly to eat our fruit or starve? If we leave a diverse flora in and around the orchard, do we reduce damage or do we provide a feeding focus and draw in additional birds? It seems by no means clear in many cases *why* pest species are attacking crops. Consideration of the breeding distribution of the bullfinch (Fig. 2) shows that the vast majority must survive perfectly well without access to pear buds, and the often sporadic occurrence of damage to pears implies a complex relationship between the bird and its various natural and man-made food resources.

Newton (1968, 1972) developed the theory that a biennial correlation existed between bud damage and the crop of ash (*Fraxinius excelsior*) seed, He argued that in years of a heavy ash crop there was little bud damage, and, conversely, in the alternating years of light ash crops, damage levels were high. More recently, in a study in the Weald of Kent, little evidence has been found of any association of damage with ash crop (Flegg & Matthews 1979, 1980). Bullfinches seem to be versatile feeders, with seeds of dock (*Rumex* spp.), bramble (*Rubus fruticosus*), and ash together with the buds of hawthorn (*Crataegus monogyna*) and blackthorn (*Prunus spinosa*) featuring largely in the diet. In addition many other plants serve as food items to a lesser degree. Even during prolonged snowfall, bullfinches were observed feeding on minute items, like dock seed or the kernels of bramble seeds, adjacent to a prolific but untouched crop of pear buds. Clearly, we need to know much more about the nutrient and other content of the various foodstuffs.

Damage is sporadic, and as yet we have no "early-warning" system that allows the prediction of appreciable damage levels. In some years, damage to an individual orchard may be very severe, while other, apparently susceptible, orchards nearby lose few buds. Thus prophylactic treatments, which in any case may be undesirable because of the problems of habituation, could be wasteful and costly, even if effective ones were available.

In temperate countries, climate has a direct influence on crop protection. Until recently protection of crops by netting has been regarded as feasible only in gardens, because of its cost. However, production techniques are reducing the cost of plastic netting and protection of fruit, such as cherries and grapes, from thrushes *Turdidae* and starlings (*Sturnus vulgaris*) may now become economic, especially with cherry, where trees of smaller stature based on the rootstock Malling Colt are now available. Problems arise in the case of netting during the winter, e.g., to protect pear buds against bullfinches, since heavy snowfall could damage both the netting and the tree beneath. Thus the grower turns

Figure 2. The breeding distribution of the bullfinch in Britain and Ireland 1968–72, plotted by 10 km National Grid squares. Large dots represent proven breeding, medium dots probable breeding, and small dots suspected breeding. Reproduced from Sharrock (1976) by permission of the British Trust for Ornithology.

his hopes to a deterrent chemical spray, but here low winter temperatures and high moisture levels pose problems to the formulation chemist. There is the risk that the repellent will be so well protected against weathering as to limit release of the active ingredient. There is an additional problem in winter that such materials have to be applied without causing undue tractor-wheel damage to the orchard soil.

Turning to tropical crops, there is good evidence that even with the blackfaced dioch (*Quelea quelea*) in Africa, cultivated cereals are not the birds' preferred food (Ward 1965; Crook & Ward 1968), as the smaller seeds of various wild grasses form the bulk of its diet.

Even in a year of severe damage, only 20% of the food intake was guinea-corn. Unfortunately, with *Quelea,* numbers more than compensate for any lack of enthusiasm for cereals, a factor compounded by the damage to standing crops caused by the sheer weight of resting birds. As with the bullfinch, a large proportion of the *Quelea* population must spend its life remote from any form of cultivation.

Climate affects the behaviour of *Quelea* in a different way. Over most of its range there is roughly a three-month rainy season followed by nine months of dry weather. The vast production of wild seed following the rains is exploited by many other birds, mammals, and especially insects. As the available stocks diminish, scattered flocks of *Quelea* coalesce in river valleys and inundation zones where seed production is greatest, to form enormous and much-feared flocks. At the onset of the next rains, the remaining seeds germinate, threatening the *Quelea* with starvation. This is the time of forced nomadic migrations to areas where the rains came earlier, and fresh seed stocks, both wild and cultivated, are at a high level. On these the ravenously destructive hordes descend.

Over much of *Quelea's* enormous range, rainfall patterns and timing are both complex and unpredictable, resulting in equally erratic nomadic migrations. Protection of crops *in situ* from *Quelea* flocks is extremely labour-intensive, even in the context of village agriculture, requiring most of the local population to turn out all day, perhaps for several weeks, in an effort to keep them away. Damage is seldom prevented, but is spread out more evenly than would occur naturally: a situation superficially similar to that advocated to control bullfinch damage with the difference that cereal crops do not have the high damage-tolerance "cushion" shown by pears.

In Britain, two cases of bird damage to cereals illustrate the necessity for ecological forethought. It was long a puzzle why rooks (*Corvus frugilegus*) were considered a pest in Scotland and as little more than a game bird in England. Feare, Dunnet & Patterson (1974) showed that the answer was not to be found in rook population densities but in the relative sowing dates of cereals in England and Scotland. In the north sowing corresponded with the breeding season and rooks satisfied their high energy and protein demands by feeding on cereal seed; in the south the early-sown grain grew away too soon so they were denied this food. More recently, maize is beginning to feature as a crop in those areas of southern England climatically suited to its growth (Feare, 1978). Relatively high soil temperatures are necessary and because of this maize must be a late-sown crop and thus has proved vulnerable to rook attack. As the maize acreage increases it will be interesting to see, whether the rook population of southern England also increases. The second example of bird damage to cereals concerns the extensive drainage over the last decade of the fresh marshes and rough grazing areas adjacent to the Thames in Kent and Essex. Much of this "improved" land is used for potato and cereal production, and in the last few years the cereals have been subject to damage by brent geese (*Branta bernicla*) (Wright & Isaacson, 1978). Light grazing of cereals during the winter can be beneficial, but damage caused by overgrazing can be considerable and compounded by the "puddling" effect on the soil caused by the feet of a dense flock of geese. It is unfortunate that the conversion of rough grazing marshland to cereals came at a time when increasing brent goose populations had put natural food resources, such as the marine grass *Zostera,* under considerable stress. Growing an acceptable alternative food almost beneath the beaks of these geese, which are coastal in distribution, and have no tradition of inland feeding, seems to be inviting damage.

Numbers of birds

To those attempting to limit the numbers of or damage done by pests operating at low population levels, the prospects of control seem much poorer than where numbers are large, and breeding, feeding and roosting are collective. A comparison of bullfinch and *Quelea* population dynamics and movements illustrates that this is not necessarily so.

A handful of bullfinches can do immense damage. Feeding at normal rates, each bird can strip half the buds from a mature pear tree in a day – during which as little as 90 minutes may be spent actually feeding on pears (Flegg & Matthews 1979). Bullfinches do not form flocks, in the normal sense, and are relatively shy, quiet and inconspicuous birds – all traits which militate against control by shooting and make live-trapping a time-consuming and expensive operation. Although only small numbers are present in any orchard at any time, these will probably form part of a larger, mobile, local population. In addition, the bullfinch is a very widely distributed bird (Fig. 2), so the "reservoir" of potential replacements for birds killed is enormous. Although pear-growing is concentrated in certain regions, notably the south-east, of the United Kingdom, even there the 3000 ha involved represent only a minute, and diffusely distributed, fraction of the countryside suitable for bullfinches. Thus assuming concerted control to be a biologically sound proposition, even in a fruit-growing district, few people would be motivated to kill bullfinches relative to the number harbouring them on their land. In the breeding season, nests are scattered and well-hidden, usually in dense and often thorny vegetation. Even when the young are large, the parents only bring food at half-hourly intervals, and commonly visit the nest together, thus minimising the risk of detection. In the circumstances it is impossible to envisage nest destruction as a practical control technique.

Studies of *Quelea* seem to present a vastly different picture. This species often congregates in enormous flocks to feed, to roost and to breed. The "weapons" used against them include dynamite and flame throwers as well as poisons, often aerially applied. Crook & Ward (1968) point out that *Quelea* are subject to varied migrations and range over $6{\cdot}5 \times 10^6$ km^2 in an area where communications are poor. Roosts vary in size, duration, and day-to-day composition. In the Republic of South Africa, despite massive kills by aerial spraying, which included over 100 million *Quelea* killed in one year, devastation of farm crops continued. Temporary respites following such expensive control measures might be locally worthwhile, but cannot be long-lived. Ward (1965) considered that *Quelea's* vast range, with so many inaccessible "reservoirs", precluded permanent reduction in numbers.

Although on a smaller scale, damage to cereal crops caused by the house sparrow (*Passer domesticus*), a relative of *Quelea,* has many similar underlying features. Control problems are compounded by the close association of house sparrows with human habitations for breeding, roosting and supplementary feeding. In these situations the dangers of poisoning, and human attitudes to wildlife destruction, are important factors to be considered.

In Britain, nationwide censuses organised by the British Trust for Ornithology indicate that, despite some year-to-year fluctuations, the populations of most breeding birds are stable; where catastrophic losses occur, as in the wren (*Troglodytes troglodytes*) during particularly severe winters, these are rapidly replenished. In simplistic terms, such stability is apparently achieved by over-producing young and leaving natural and artificial mortality factors, presumably largely density-dependent, to do the rest.

Much argument surrounds the role of the predator, be it a hawk or a man with a gun, in population dynamics. Most natural predators are not prey-specific but feed on what can most profitably be caught; it is now axiomatic that in normal circumstances predator numbers depend on prey availability and not *vice versa*. Thus the introduction of sparrowhawks (*Accipiter nisus*) would not be expected to have measurable impact on bullfinch numbers, nor would encouragement of nest predators, such as jays (*Garrulus glandarius*) be expected to materially reduce nesting success.

Murton (1965, 1968, 1971) and others argue that predators take only part of the natural over-production, and base their reasoning on the view that causes of death are compensatory rather than additive. So, man as a predator only shoots part of a surplus that was destined to die before the next breeding season anyway. Murton suggests that the early removal of wood-pigeons (*Columba palumbus*) from the population merely allows the remainder to feed well and survive better, and argued that shooting, though offering some protection to individual crops, functioned more to scare birds than limit their populations. This begs, to a degree, the important questions of *when* and *why* natural mortality occurs, a subject on which we urgently need more information. Much depends on the temporal pattern of occurrence of peak natural mortality, peak pest numbers and damage. Peak numbers are usually thought to occur at the end of the breeding season, and heavy damage might in some circumstances be avoided by planting early-maturing cultivars. If crop plants were only incidental dietary constitutents such a strategy would work, but species having considerable dependence on crops, like the house sparrow, might be expected to compensate for the change in due course by some advancement of the breeding season.

If crop damage occurs during the winter, then the pattern of natural mortality becomes more important. Damage could be avoided, or alleviated, if the vulnerable crop became

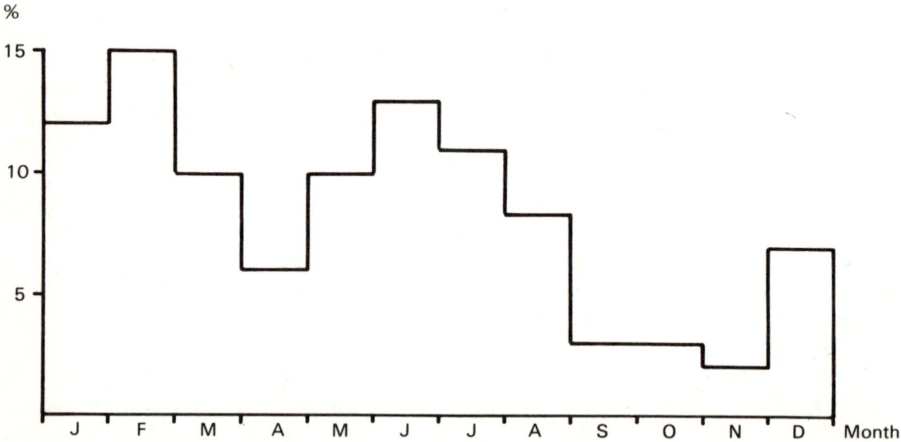

Figure 3. Percentage bullfinch mortality by months derived from recoveries of ringed birds. Redrawn from Mead, C. J. (1974) *Bird Ringing* by permission of the British Trust for Ornithology.

available only after natural mortality had considerably reduced the population. This is not so with the bullfinch, where a preliminary assessment (Fig. 3) based on recoveries of marked birds – indicates a peak in mortality late in the winter and another in midsummer. Here is a case where early removal of the population surplus, an "artificial" mortality caused by trapping or shooting, may avert damage to local crops. Again, much more information is needed, to allow measurement of the speed and degree of subsequent infilling from the surrounding population. Damage by birds is additive, and perhaps shooting and trapping are too often dismissed as damage-reducing measures on theoretical grounds, without adequate information on the potential savings. Each situation should be assessed before any control strategy is adopted.

Migration

One other important natural factor remains to be discussed: the seasonal migratory movements of birds, to which should be added those irregular movements provoked by sudden climatic changes. In reviewing the movements of species associated with damage to crops, Flegg (1974) calculated the distances involved from recoveries of ringed birds in Britain (Table 1). For most species, although distances of movement are not great, they are clearly adequate to allow infilling after their numbers have been reduced by control measures. Given the flexibility of behaviour shown by many species which move away (usually south or west) from periods of severe weather and return promptly afterwards, it seems fair to assume that infilling will occur if the population size is adequate. If severe weather causes heavy mortality then infilling is likely to be slight and slow.

Table 1

Distances of movement of some species associated with damage to agricultural crops as shown by ringing recoveries. The proportion of recoveries in each of three distance categories, and the sample sizes, are tabulated under two arbitrarily chosen periods: summer (April-August inclusive) and winter (September-March)

	Summer				Winter			
		Distance, km				Distance, km		
	No. recovered	0–10 %	11–100 %	>100 %	No. recovered	0–10 %	11–100 %	>100 %
Wood-pigeon	118	73	26	1	339	68	30	2
Collard Dove	50	64	16	20	128	63	13	25
Skylark	29	72	21	7	16	63	13	25
Rook	119	76	24	0	157	66	33	1
Great Tit	320	93	6	< 1	460	87	12	1
Blue Tit	449	93	7	< 1	1602	86	13	1
Bullfinch	208	92	7	< 1	346	88	12	< 1
House Sparrow	656	95	5	< 1	993	94	5	< 1

Basic annual migrations provide some problems when considering control strategies. Are the flocks of migrant starlings, so obvious during the winter, the birds that cause agricultural problems? If concern is over damage to ripening cherries, then the answer is "no", but if the problem is damage to feedstuffs at intensive animal husbandry units (Feare 1975) then the answer may be "yes". Hence measures aimed at population control, rather than physical protection against damage, need careful planning, or funds and effort may be wasted in eliminating birds of non-pest status.

There has been considerable argument over the role of migrating birds as vectors of disease. Snow (1968) and Flegg & Mead (1975) detailed the convincing evidence that starlings do not transmit foot-and-mouth disease in cattle. More recently Beer (1979) has suggested that the feeding and roosting habits of migratory starlings are consistent with their having spread fireblight disease of pears (caused by the bacterium *Erwinia amylovora*) from Britain to northern Europe. Radar and direct observations show that some movement of birds may occur at almost any time during the year, so the fact that the bulk of migration of starlings across Europe occurs well before the likely infection period, at blossom time, may not be wholly relevant. Nevertheless, so many other factors e.g., insect movements, wind direction, weather conditions, are involved that an actual demonstration of the vector role is necessary. Should starlings prove to be responsible for spreading the disease, then control measures against them will be extremely difficult to devise. The number of birds involved is prodigious; their origins, destinations and detailed patterns and timing of movements are difficult to predict; and, of course, only a minute proportion may carry the disease. In this case, the cultivation of resistant or tolerant varieties, if they exist, seems a far more profitable defence, as it would safeguard the fruit grower no matter what the mechanism of transmission.

"Irruptive" movements, when normally sedentary species suddenly migrate under pressure from high populations or low food supplies, present a further hazard. The great tit (*Parus major*) and blue tit (*P. caeruleus*) both cause some damage to fruit and are subject to occasional, and unpredictable, irruptive movements, usually westward from Europe to Britain. Wright (1959) recorded unusually severe attacks by these species on ripening fruit associated with an irruption on a huge scale in 1957–58.

The sudden arrival, and highly successful exploitation of an apparently vacant "niche", of an alien species poses instant problems when control strategies have to be devised. The collared dove (*Streptopelia decaocto*) is one such. It is thought that a sudden genetic change in this predominantly Asian species led to a rapid westward expansion across Europe – so rapid that Britain and Ireland were colonised, and in parts overrun, in a decade. The species is particularly troublesome to poultry farmers and as a pest in grain stores. A different example is provided by the ring-necked parakeet (*Psittacula krameri*), in south-eastern England. Here an increasing feral population, derived from escaped pets, breeds effectively and is causing damage to ripening fruit, especially apples. It was thought that winter survival would prove the limiting factor, but the parakeet has shown itself well able to exploit human kindness by feeding at suburban bird-tables. Any control strategy in this case will need to take into account the problem of public reaction to the destruction of a colourful bird.

Behaviour

In seeking means of repelling or deterring birds from damaging crops, a new biological factor enters our thinking in devising control strategies. Whether the repellent/deterrent operates on birds' physical senses, or *via* audio, visual or chemical stimuli, we must remember that we are dealing with creatures that are usually highly adaptable and which possess versatile patterns of behaviour. It may be that they are eating crops in order to survive at a critical time: a motivation that merely distasteful deterrents are unlikely to counter.

Against this background, it can be argued that in-flight obstacles such as black "cotton", or visual deterrents like scarecrows, or flapping yellow ribbons, are unlikely to be a lasting success. Even model hawks, suspended from gas-filled balloons, will not function as scarers unless they can be provided from time to time with "teeth" which confront birds with actual danger. The basis for a new generation of scaring devices may be found in a study of plumage patterns and their use in communicating alarm: the white bars visible on the outstretched wings of a wood-pigeon serve this purpose. Even here some reinforcement of the stimulus may soon prove to be necessary, in much the same way as the monarch butterfly (*Danaos plexippus*) backs up its warning colouration with a body chemistry that is an effective emetic.

The problems associated with chemical repellents are broadly similar. Flegg *et al.* (1977) pointed out the scarcity of candidate repellent materials, and outlined the problems surrounding their use. An important consideration is the need for effective materials to resist weathering under field conditions. The bullfinch poses an additional problem: in feeding, buds are picked more or less whole and then husked by rapid and dextrous mandibulatory movements; bud scales and leaf initials fall to the ground and only the flower initials are eaten. Thus, little of any treatment applied to the bud is in contact with the tongue or palate for long. Several promising deterrent materials are irritants, and may not be suitable for health reasons; others, which function by conditioned aversion, may have such effective "teeth" as to be unacceptable to either the Pesticide Safety Precautions Scheme or the general public.

Conclusions

Most birds associated with crop damage are too numerous and too widespread for their populations to be estimated with accuracy. They have considerable reproductive potential, often amounting to routine over-production of young and possess, to varying degrees, an ability to travel – the distance and timing depending sometimes on evolutionary history and often on more immediate, less predictable, factors such as food supply or weather conditions.

The combination of high reproductive capacity, catholic diet, and mobility is sufficient to indicate that rarely, if ever, will an avian pest be eliminated entirely, or even have its population materially reduced. For long- and medium-haul migrants, some prevention of damage, however difficult, is likely to be more easily achieved than a sufficient population reduction. In the case of starling damage to sweet cherries, fruit breeders at East Malling have now produced a rootstock, Colt, that gives a tree small enough to be netted and the cost of such protection is being reduced by new techniques of net-production. There are numerous sad examples, typified by attempts to control *Quelea* populations, of the cost and ineffectiveness of destruction tactics. For short-haul migrants, like the bullfinch, local control at an appropriate time of year may prove effective, but there is urgent need for more evidence on the speed and degree of infilling that follows. Live-trapping is practised by many fruit growers, but with very variable results: there is scope for research on the most effective timing, and on the mechanisms by which call-birds or baits operate. Our lack of detailed understanding of birds behavioural response is highlighted by the general simplicity of most scaring devices, and by their short-lived effects. An understanding of behaviour plays a part, too, in the assessment of materials as potential repellents and in the search for resistant varieties. Such work may be fruitless until we know how and why birds choose their food.

There is also an urgent need for much more ecological information on the dietary preferences of pest species. What nutritional or other elements do they derive from the components of their diet? Are there alternatives? By the ruthlessness of our crop husbandry, are we sometimes forcing birds to eat our crops or starve? Before new crops, or crop-production techniques, are introduced, realistic appraisal should be made of the possible costs of protecting the new system from bird damage, and of the cost of that damage if unabated. Too often we see situations where man has stepped in to provide a crop as food when natural supplies are short: millet for *Quelea,* maize for rooks, winter-cereals for brent geese and pear trees for bullfinches are cases in point. For many pests, a judicious reduction of risk, based on ecological studies, may ultimately be coupled to an integrated spectrum of crop-protective measures to minimise damage. In the long run, such cost-effective studies could produce a cheaper solution to crop-protection problems that the present blunt-instrument tactics so often resorted to in an attempt to reduce damage after the event.

References

BEER, S. V. (1979) Fireblight inoculum: sources and dissemination. *EPPO Bulletin* 9, 13–25.

CROOK, J. H.; WARD, P. (1968) The Quelea problem in Africa. In: Murton, R. K.; Wright, E. N. (*Eds*) The problems of birds as pests, p. 211–229. London; Academic Press.

FEARE, C. J. (1975) Cost of starling damage at an intensive animal husbandary unit. *Proceedings 8th British Insecticide and Fungicide Conference 1975*. p. 253–259.

FEARE, C. J. (1978) The ecology of damage by rooks (*Corvus frugilegus*). *Annals of Applied Biology* 88, 329–334.

FEARE, C. J.; DUNNET, G. M.; PATTERSON, I. J. (1974) Ecological studies of the rook (*Corvus frugilegus* L.) in north-east Scotland. Food intake and feeding behaviour. *Journal of Applied Ecology* 11, 867–896.

FLEGG, J. J. M. (1974) Seasonal movements of birds. In: Bird damage to crops and the background to control. *Annals of Applied Biology* 76, 341–345.

FLEGG, J. J. M.; MATTHEWS, N. J. (1979) Bullfinch damage. *Report of East Malling Research Station for 1978 (1979),* 128–129.

FLEGG, J. J. M.; MATTHEWS, N. J. (1980) Bullfinch damage. *Report of East Malling Research Station for 1979 (1980),* in press.

FLEGG, J. J. M.; MEAD, C. J. (1975) Foot-and-mouth disease and migrant birds. *Veterinary Drug,* February 1975, 3.

FLEGG, J. J. M.; MUIR, R. C.; HUNTER, L. D. (1977) Preliminary tests of chemical repellents to bullfinches. *Proceedings 1977 British Crop Protection Conference – Pest and Diseases* p. 469–475.

LEWIS, M. F. (1967) Associate Committee on Bird Hazards to Aircraft. *Bulletin No. 5 National Research Council, Ottawa, Canada.*

MURTON, R. K. (1965) The Wood Pigeon, London; Collins.

MURTON, R. K. (1968) Some predator-prey relationships in bird damage and population control. In: Murton, R. K.; Wright, E. N. (*Eds*) The problems of birds as pests, p. 157–169. London; Academic Press.

MURTON, R. K. (1971) Man and Birds. London; Collins.

NEWTON, I. (1968) Bullfinches and fruit buds. In: Murton, R. K.; Wright, E. N. (*Eds*) The problems of birds as pests, p. 199–209. London; Academic Press.

NEWTON, I. (1972) Finches. London; Collins.

SHARROCK, J. T. R. (1976) The Atlas of Breeding Birds in Britain and Ireland. Berkhamsted; Poyser.

SMITH, B. D.; KENDALL, D. A. (1975) The economic significance of different levels of bird damage to plum and pear buds. *Proceedings 8th British Insecticide and Fungicide Conference (1975)* p. 245–251.

SNOW, D. W. (1968) Birds and the 1967–68 foot-and-mouth epidemic. *Bird Study* 15, 184–190.

STENHOUSE, D. (1962) A new habit of the redpoll *Carduelis flammea* in New Zealand. *Ibis* 104, 250–252.

SUMMERS, D. D. B.; POLLOCK, M. R. (1978) The effects of bullfinch damage on the yield of pear trees. *Annals of Applied Biology* 88, 342–345.

WARD, P. (1965) Feeding ecology of the black-faced dioch *Quelea quelea* in Nigeria. *Ibis* 107, 173–214.

WRIGHT, E. N. (1959) Bird damage to horticultural crops. *Journal of the Royal Horticultural Society* 84, 426–434.

WRIGHT, E. N.; ISAACSON, A. J. (1978) Goose damage to agricultural crops in England. *Annals of Applied Biology* 88, 334–338.

Waterfowl crop damage prevention and compensation programs in the Canadian Prairie Provinces

H. Boyd
Canadian Wildlife Service
Ottawa

Summary

Damage to ripening cereal crops by ducks and geese has become widespread in the Prairie Provinces in the past 30 years, being greatest when harvesting-weather is wet. Since 1972, the federal government has shared the costs of provincial damage prevention and compensation programs. There are some 94 areas where damage is frequent and can be minimised most cost-effectively by annual preventive activities. Scattered and sporadic losses elsewhere are better dealt with by compensation, preferably through an insurance scheme. Swathing, introduced to deal with uneven ripening, invites duck damage and should be eliminated where possible. Non-susceptible crops should be grown near large lakes where ducks are abundant. Compensation payment arrangements must not be open-ended and should encourage farmers to avoid waterfowl damage. Though who should pay for damage, and how, are subjects of controversy, farmers should not be expected to suffer substantial financial losses for the benefit of distant users of waterfowl.

Introduction

The wildlife damage prevention and compensation programs of the provinces of Alberta, Saskatchewan and Manitoba, intended to reduce financial losses to farmers resulting from crop damage by wildlife, were extended in 1972 by the signature of federal-provincial agreements for shared funding of programs to minimise losses to grain crops inflicted by waterfowl (ducks, geese and cranes). The experience gained in these large-scale programs may help in the design of effective programs elsewhere. Cost-effective damage prevention schemes have been developed to deal with areas where damage is reported in most years. Much of the damage due to ducks is widely scattered and affects most individual farmers only infrequently. Such scattered losses could be handled most economically by an insurance scheme. The choice of damage prevention and compensation schemes has come to be dominated by political, rather than by technical or financial, considerations. That could well happen in other countries too.

Sugden (1976) has provided a thorough review of the nature and extent of damage by waterfowl to barley and wheat crops and of the ways of minimising that damage, as well as outlining the needs for additional research. I will deal very briefly with the historical picture and concentrate on some very recent events, which have caused significant changes in planning programs for 1979 and beyond.

History

Because the prairie provinces are full of wetlands, from large lakes to millions of small sloughs and potholes, more or less temporary, they have long been the home of the greatest concentrations of breeding ducks in North America. They are also used by many geese on migration in the autumn. The chief grain-growing areas of Canada thus coincide with the most important waterfowl areas. Though this led to some attacks by ducks on crops from the beginning of cereal farming in the former grasslands and parklands, less than a century ago, damage to grain by field-feeding ducks, geese and cranes did not become substantial until about 30 years ago. At that time the practice of swathing – cutting the crops of barley and wheat and allowing them to lie in rows on the ground for 10 days or more before picking-up by combine harvesters – was introduced in order to bring about more even ripening of the crop than would otherwise be achieved. Though swathing is unnecessary in the drier parts of Alberta it is very widely practised, especially in the northern cereal-growing areas where the risk of failure to ripen is considerable, due to the shortness of the growing season and the early onset of cold weather.

Swathed cereals are more easily available to field-feeding waterfowl than still-growing crops or stooked sheaves. This has led to extensive feeding on barley and wheat by mallard (*Anas platyrhynchos*), pintail (*A. acuta*), white-fronted geese (*Anser albifrons*), lesser snow geese (*A. caerulescens*), Canada geese (*Branta canadensis*) and sandhill cranes (*Grus americana*). The amount and extent of damage are determined chiefly by the length of time the crop lies on the ground, which in turn depends chiefly on the amount of rain during the harvest period. In dry years an individual crop may need to lie little more than a week. In wet years it may not be picked up for three weeks or more and may deteriorate greatly in quality and value as a result. The start and duration of harvesting vary regionally, as well as from year to year. In such northern areas as the Peace River district of Alberta successful harvests are the exception rather than the rule; grain-growing there would probably not be continued were it not for government subsidies.

Because of the importance of farming in the economy and politics of the prairie provinces, all three provincial governments have long had programs to compensate farmers for damage to crops by wildlife, including large mammals such as moose, elk and brown bears, though not mice and voles, as well as birds. As the costs of these programs rose, and as the extent of damage by waterfowl tended to increase, the provincial governments turned to the federal government for financial assistance. They argued that, although wildlife are a provincial responsiblity, the federal government has legislative responsibility for migratory birds, in the form of the Migratory Birds Convention Act, which followed the signature in 1916 of the Migratory Birds Convention between the U.S.A. and Great Britain on behalf of Canada. Although the Convention makes provision

Table 1

Areas of grain crops in swath and vulnerable to waterfowl damage in Saskatchewan during 1978 harvest: in millions of acres.

	Spring Wheat	Durum	Barley	Oats	Total cereals	% Area sown
Area sown	14·2	2·9	3·0	1·2	20·3	
Area in swath						
21 August	3·8	1·0	1·0	0·3	6·1	30·2
28 August	7·1	1·6	1·3	0·5	10·5	51·5
11 September	5·4	0·8	1·1	0·4	7·6	37·4
18 September	4·8	0·7	1·1	0·4	7·1	34·8
25 September	3·5	0·6	0·8	0·3	5·6	27·7
2 October	1·8	0·1	0·4	–	2·3	11·6

for protection to be removed from species shown to be causing agricultural damage, there is nothing in the Convention or the Act obliging the federal government to assume liability for damage caused by migratory birds. Yet in 1972 the federal government entered into agreements with the provincial governments to share equally the costs of waterfowl damage prevention and compensation programs, subject to upper limits on total funding and to agreements on the allocation of the available funds between the three provinces.

Table 2
Federal expenditures on waterfowl depredation programs in the prairie provinces in crop years 1972 to 1978. Provincial expenditures matched or exceeded the federal contributions. In thousands of dollars, rounded to nearest 0·1.

Year	Province	Prevention	Compensation	Total
1972	Alberta	54·8	200·0	254·8
	Saskatchewan	24·4	250·0	274·4
	Manitoba	49·0	3·9	53·0
	Total	128·2	453·9	582·2
1973	Alberta	107·2	433·5	540·7
	Saskatchewan	24·0	281·2	305·2
	Manitoba	50·0	30·8	80·8
	Total	181·1	745·5	926·7
1974	Alberta	176·4	441·8	618·2
	Saskatchewan	174·7	285·1	459·8
	Manitoba	172·1	89·0	261·1
1975	Alberta	200·3	244·4	444·8
	Saskatchewan	216·0	114·4	330·4
	Manitoba	83·1	141·7	224·8
	Total	499·4	500·6	1000·0
1976	Alberta	213·4	190·4	403·8
	Saskatchewan	208·8	97·5	306·3
	Manitoba	53·9	61·6	115·5
	Total	476·1	349·5	825·6
1977	Alberta	243·0	565·6	808·6
	Saskatchewan	296·5	165·3	461·7
	Manitoba	65·9	163·8	229·7
	Total	605·4	894·6	1500·0
1978	Alberta	402·3	600·0	1002·3
	Saskatchewan	315·9	675·0	990·9
	Manitoba	181·7	225·0	406·7
	Total	899·9	1500·0	2399·9

The areas of cereal crops susceptible to waterfowl damage are large (Table 1) and damage is widespread. Table 2 demonstrates that substantial sums have been spent on the programs. The original agreements expired in 1977. Despite provincial pressure for further long-term agreements, with increased funding to offset both the rising costs of grain-growing and the spread of reported waterfowl damage, the federal government agreed only to an extension for one year in 1977 and again in 1978. The hostility created by these one-year agreements, in the farming community as well as in provincial administration, was heightened in September 1978 when, as part of a federal cost-cutting campaign, it was announced that the federal contribution to waterfowl crop damage programs would be discontinued after 31 March 1979.

Damage prevention

The earliest form of damage prevention consisted of scaring waterfowl off crops they had started to attack or discouraging them from settling, by using scaring devices, accompanied by occasional gunshots. Acetylene exploders were once used but these have now largely been replaced by propane-powered exploders, of which about 3500 are in use in Saskatchewan. The government agencies have offered advice on scaring techniques and provided small numbers of exploders for demonstrations; but scaring is seen as the task of individual farmers. It is, in any event, of doubtful value as a measure of reducing total damage even if effective in frightening birds off any one field; usually the scared birds merely go to feed in another field.

Once ducks have started using a swathed crop, it is more economical to encourage them to remain there than to scatter them around, unless the particular crop is of especially high quality. Realization of this fact led to the use of "lure crops" or "feeding stations" to hold birds on limited areas. Where damage occurs in most years it is sensible to grow lure crops, primarily intended for use by birds, or to buy a crop from a farmer as soon as it has been attacked. In other places it may be more effective to lay out threshed grain on the shores of major waterfowl resting areas.

Table 3
Costs of waterfowl crop damage prevention programs in the prairie provinces, 1978

Province	No. of control units	Grain	Scaring activities	Operations and equipment	Provincial salaries and administration	Total
		$	$	$	$	$
Alberta	39	377,800	208,885	66,505	151,319	804,509
Saskatchewan	33	351,800	120,000	39,900	120,170	631,870
Manitoba	20	168,800	92,000	20,400	82,100	363,300
Total	92	898,400	420,865	126,805	353,589	1,799,479

Extensive investigations since 1972 have established how to get good value for the funds and effort expended on deterrence and on holding birds at bait stations or lure crops. At least 94 sites have been identified (39 in Alberta, 33 in Saskatchewan and 22 in Manitoba) where potential damage is so frequent and heavy as to justify preventive actions. In 1978 increased funding made it possible to operate in nearly all those locations. Table 3 shows the distribution of costs between provinces and between different facets of the operation in that year. The average cost per site was around $20 000, including a prorated amount of general administrative costs, with operating costs ranging from $7 700 to $49 300 per site. The most expensive operations were in Manitoba (Carrot Valley and

Oak Hammond Marsh) and involved areas where geese were numerous and heavily involved in field feeding. Other expensive operations in Saskatchewan also involved geese. Presumably their larger per capita consumption of grain adds to the costs.

Compensation

When the federal-provincial agreements were signed in 1972 the federal Cabinet indicated that during the life of the agreements attempts should be made to treat waterfowl damage to crops as an insurable risk. Saskatchewan already did that, but neither Alberta nor Manitoba was prepared to do so. Provincial and federal agricultural officials argued that: (1) the non-random distribution of damage, both geographically and temporally, made it actuarially unsuitable for insurance; and (2) farmers should not be expected to pay premiums to protect themselves against a risk due to the activities of "the Queen's ducks". Neither argument stands up to close examination, but they have prevailed so far.

A study of waterfowl damage claims paid in Alberta for the eleven crop years 1964–74 showed that payments had been made for claims relating to 7401 sections of land (a section is 1 sq. mi.) widely scattered throughout the grain-growing region of the province: 76·6% of those sections had claims in only one of the eleven years; 16·2% in two years; 4·8% in three years; and only 6·4% in four or more years. Such a frequency distribution suggests that the objections to insurance on grounds of non-randomness may be ill-founded.

The Saskatchewan Government used an insurance scheme for dealing with waterfowl crop damage from 1953 to 1977. Farmers had to enrol by 10 August to qualify for assistance and to pay a 2% premium. The necessary revenue was supplemented by the yield from a $1.00 impost on provincial game bird, antelope and deer hunting licences. The scheme was self-supporting until 1968–69, when extremely heavy depredation bankrupted the fund. Thereafter it had also to be supported by general revenue. From 1956 to 1967 the average number of policies taken out was only 424 and the average number of claims made 194. From 1968 to 1977 the average number of policies was 1140 and of claims 468 (ranging from 371 in 1976 to 791 in 1971); that is, about half the people taking out insurance made claims, ranging from only 29% in 1976 to 67% in 1968. Up to 1967 the largest total annual payment was $161 800. In 1968 the total cost rose to $398 900 and the average for the decade 1968–1977 was $454 400, with a peak of $503 100 in 1974. The average amount paid for an individual claim was $448 in 1956–1967 and $813 in 1968–1977; there was an upper limit of $25 an acre on the amounts payable.

From a comparison of the number of policies taken out with the amount of damage subsequently reported it appears that, particularly after 1968, farmers were quite successful in guessing whether they would suffer damage, except in 1976. The total amounts paid in claims were less simply related to the number of policies taken out. This does not imply that the farmers were unsuccessful in estimating how much damage they would suffer: that they were not required to do. Because of the $25 an acre maximum, the insurance program paid out between 42% (in 1977, a wet harvest year) and 67% (in 1976, a dry harvest) of the total estimated damage on claims. Fluctuating crop prices as well as severity of depredation affected the discrepancies. High world grain prices in 1973–1975 increased the estimated damage and the difference between damage and compensation.

For the 1978 harvest the Saskatchewan Government decided to drop insurance and introduce a compensation program, with maximum payments up to $50 an acre. This decision related less to deficiencies in the insurance scheme than to a determination to join the other provinces in emphasising the alleged federal responsiblity for damage by waterfowl. It led to a great change in the number and pattern of damage claims. Although the extent of damage in 1978 was believed to be only slightly above average and certainly below that of 1977, the number of claims submitted to the Saskatchewan Crop Insurance Board, which remained responsible for program administration and field adjusting, rose to 1835, the highest yet recorded, having been approached only in 1971 (1637 claims). The total calculated compensation due in 1978 was $1 836 500, 57% of the total estimated damage. Actual payment was limited to $1 500 000, because of insufficient federal and provincial funds, at an average rate of $33.50 an acre.

Not only did the number of claims filed in 1978 show a great rise, there were also marked changes in the distribution of claims. About 3·4% of the grain farmers in the province filed claims, representing damage to about 0·2% of the crop acreage. Claimed damage was greater in the north (0·4% of acreage) than in the south (0·07% of acreage). There was a surge in the number of claims from the northeast of the province. The north is an area of mixed farming, with less of the farmed land being devoted to cereals than in the south. The damage prevention program was largely concentrated in the south.

The value of the average claim rose to $1 000 in 1978; eleven claims were between $5 000 and $10 000 and three exceeded $10 000, one being just under $23 000. This increase appeared to be chiefly a result of raising the limit to $50 an acre. For claims in the range $500–$2 500, the $50/acre maximum was penalizing the farmer with good crops and paying off in full for poor crops.

The most dramatic change in the compensation data was the number of townships with claims in 1978 that had no previous history of claims; in 55% of cases no claims had been filed in the previous 13 years. Presumably, in addition to normal turnover, the compensation program induced many farmers to file claims that they would previously have considered marginal for insurance.

The departure of the pattern of claims in 1978 from that in earlier years raises questions about the potential increase in claims to be expected should Saskatchewan adhere to a compensation program, though the 1978 results were sufficiently alarming to cause the provincial government to announce its intention to return to insurance in 1979.

Apart from the tendency of compensation programs to obey Parkinson's laws, with the demand expanding to meet and exceed the budgeted funds, insurance is preferable to compensation because it can provide incentives to long-term changes in land-use and agriculture technology. If provincial arguments that premium payments are too much of a burden for individual farmers are acknowledged a mixed program might be considered: compensation at, say, $25 an acre, with insurance coverage available above that amount. Such a mixed program would reduce the total cost to the taxpayers, while retaining the merits of insurance, even if somewhat diluted.

A waterfowl damage insurance program should have amongst its aims the encouragement of: (1) growing crops that do not attract waterfowl; (2) avoiding planting of vulnerable crops in vulnerable locations; (3) alternatives to swathing that are less attractive to waterfowl; and (4) reduced costs to farmers and governments. In its wildlife damage compensation program the Alberta Government has introduced a reducing scale of payments that provides an incentive to desist from growing crops in vulnerable sites: payment of one claim on an area in five years is at the rate of 80% of value; a second claim in five years is paid at 70% of value, a third at 60%, and so on.

Proposed future programs

The federal decision of September 1978 to withdraw support for damage prevention and compensation programs after 31 March 1979 was followed by strong protests from the provincial governments, from farmers' organizations and from some hunters' and naturalists' organisations. In representing the case for continued and improved programs the federal Departments of Environment and Agriculture proposed that the fund for compensation be increased, to a maximum of $3 million of federal money, to be matched by provincial funds. The proposed damage prevention program included a new element, the suggestion that in 14 areas where large-scale waterfowl damage was chronic attempts should be made to buy substantial tracts of land and manage them primarily for waterfowl, an approach that has been used successfully at Last Mountain Lake, Saskatchewan. The land would be purchased and managed jointly, with the title going to the provincial government.

At the time this paper was delivered the decision of the Federal Cabinet on this program submission had not been made. The Cabinet subsequently accepted the recommendation for restoring and enlarging the damage prevention and compensation programs. It directed that up to $3 million of federal funds be spent in each of the next five years on compensation and that up to $1·3 million be spent on damage prevention. Of the $1·3 million up to $600 000 could be spent on prevention activities of the kinds previously in use while up to $700 000 could be spent on acquiring land in districts where heavy damage is chronic. Though this decision provides a basis for fresh negotiations with the provincial governments, it is too soon to tell whether any new agreements will include the federal proposals in anything like the form proposed.

The nature of the activities to be undertaken and the scale of funding that may be appropriate are the subject of political decision, in which agricultural considerations predominate over wildlife ones. As a contribution to food production, the programs must be designed to minimise losses of harvestable grains, especially those of high quality.

A policy decision to continue the programs having been made, it becomes the task of the program managers in the provincial and federal governments to design and implement a cost-effective collection of projects. This will involve continuing appraisal of each element in the programs, because the relation between costs and benefits is highly variable from site to site and year to year. There is no reason to expect that an optimum mix can be achieved but with sufficient vigilance it should be possible to avoid too great a waste of public funds.

Acknowledgements

Federal and provincial collaboration in the conduct of waterfowl damage prevention and compensation programs has involved many players in the gathering and assimilation of data, as well as in practical activities in the field. All their contributions have helped in the development of the programs. For the specific purpose of preparing this paper I am especially indebted to H. J. Poston, D. Jurick and Dr. L. G. Sugden, Western and

Northern Region, Canadian Wildlife Service, for data, analyses and for the slides used to illustrate the oral presentation; to J. O. Jacobson and R. K. Schmidt, Jr., of Interdisciplinary Systems Ltd., Winnipeg, Manitoba, for an evaluation report on the 1978 crop damage program in Saskatchewan; to Dr. V. E. F. Solman, CWS, Ottawa; and to G. M. Gorrell, Director, Crop Insurance, Food Production and Marketing Branch, Agriculture Canada, Ottawa.

Reference

SUGDEN, L. G. (1976) Waterfowl damage to Canadian grain: current problem and research needs. *Canadian Wildlife Service Occasional Paper 24,* 1–24.

Bird damage and the law

A. W. Colling

Summary

British bird protection legislation and those of its provisions currently affecting crop protection are discussed in the context of the Council of Europe Directive on the Conservation of Wild Birds, adopted on 2 April 1979. National legislation will be required to conform to the Directive's provisions and, although this will necessitate only minor alterations to British laws, the amending legislation must be enacted by April 1981. However, interested parties will undoubtedly take advantage of this opportunity to propose additional amendments, possibly more far-reaching than those required by the EEC, to place further restrictions on legitimate control of pest species.

Conflicting views between growers and bird protectionists must be resolved and a plea is made for the exercise of strict objectivity by both factions in judging whether an individual pest species needs statutory protection and, if so, in what circumstances and to what extent.

It is emphasized that all bird populations are ultimately limited by the carrying capacity of habitat and environment. Further extension of legal protection cannot alter these limits appreciably, but might well be counter-productive in alienating the farmers' goodwill for wildlife conservation and in reducing the very real contribution to positive environmental management which he, more than anyone else, can make.

Introduction

Before 1954, bird protection legislation in force in Britain included at least nine major Acts of Parliament, dating from 1880 to 1933. These Acts contained national regulations and, in addition, there were a large number of local Orders which applied mainly on a county basis. Public awareness of these laws was largely limited to the contents of a notice displayed outside police stations. This notice listed the names of those species which the local County Council had decided to protect by making it an offence to take or kill the birds or steal their eggs. Harmful birds were not included in these lists and only brief mention was made of game birds and species that are the traditional quarry of wildfowlers. With so many separate Acts, and lists of protected birds differing from county to county, it is not surprising that very few people, including the police, understood the bird protection regulations; consequently enforcement was ineffective and abuse of the law by egg and bird collectors was rife. On the other hand, the control of pest species was less restricted than it is today.

The Protection of Birds Act, 1954

In 1954 Parliament enacted new legislation on bird protection and this, together with the amending Act of 1967, did much to reduce the confusion. Firstly, it repealed all the older Acts; secondly, it adopted a comprehensive system, starting with the premise that *all* wild birds are protected, the only exceptions being those that are acknowledged pests and species traditionally shot for sport. The pests and quarry species are, of course, specified and since they are, or should be, quite familiar to those who have the right to kill them, this comprehensive system has the advantage that neither the specialist nor the less-well-informed man-in-the-street can any longer plead ignorance of the law. It must also have made the policeman's job easier. The 50 or so rarest British breeding birds which are also specified in a schedule to the Act, had been particularly harried hitherto by egg collectors; for these species the 1954 Act provided *special* protection, with heavy penalties for offences. So the main provisions of our current law are:

(1) Specified pest species (Second Schedule) may be taken or killed by authorised persons* at any time; certain methods of taking or killing being prohibited.
(2) Quarry species (Third Schedule), like game birds, may be shot during the prescribed open season, but are protected during the breeding season.
(3) All other species are protected throughout the year.
(4) The rarest species (First Schedule) are given special protection i.e., extra penalties for killing or taking birds or eggs.

The birds with which we are mainly concerned at this meeting are those listed on the Second Schedule to the Protection of Birds Act 1954 and which, at that time, were regarded as harmful to fisheries, game bird and poultry rearing, sheep farming, grass and arable crops and fruit. The list does not include greylag (*Anser anser*), pinkfooted (*Anser brachyrhynchus*) and brent (*Branta bernicla*) geese, whooper swans (*Cygnus cygnus*), mute swans (*Cygnus olor*) and skylarks (*Alauda arvensis*), which are protected species in England and Wales, despite the fact that nowadays they cause concern during late winter and spring by grazing pastures, cereals and sugar-beet seedlings. (N.B. greylag and pinkfooted geese may be shot during the open season). So, although the farmer may at any time freely shoot woodpigeons (*Columba palumbus*), stock doves (*Columba oenas*), feral pigeons (*Columba livia*), crows (*Corvus corone*), rooks (*Corvus frugilegus*), jackdaws (*C. monedula*), jays (*Garrulus glandarius*), magpies (*Pica pica*), starlings (*Sturnus vulgaris*), and house sparrows (*Passer domesticus*) he may be taken to court if he kills any protected species and may be convicted and fined unless he can prove that he shot them while they were in the act of causing serious damage to crops or other forms of property (Section 4(2)(a) of the 1954 Act). Section 5 of the 1954 Act lists prohibited methods of killing or taking birds, offences being subject to a special penalty. For practical purposes, a farmer wishing to protect his crops against depredating birds – even though they be listed as harmful – may only shoot or use a cage trap. The use of any other trap or snare, or of explosives, nets, gas, poisons or stupefying substances or electrocuting devices is prohibited. While shooting and cage trapping have their uses, they are obviously unsuitable or uneconomic in many cases. So, with this limitation, together with the requirement of Section 4(2)(a) that the farmer must carry the onus of proof to justify in court his killing of any protected species, it cannot be denied that agricultural pest control is poorly served by this legislation. Section 10(1)(d) of the 1954 Act, however, empowers the Minister of Agriculture (or in Scotland, the Secretary of State) to grant a licence to any person to use poisoned or stupefying baits, subject to compliance with specified conditions, to take or kill species listed on the Second Schedule. This sounds useful in theory, and it can be very effective, but the requirement for close supervision and selective use, to avoid any risk to protected birds or other wildlife, imposes severe limitations; therefore, in practice, this method tends to be used mostly in enclosed premises, and seldom in the field.

*Broadly defined as the owner, occupier or person shown to have been authorised by the owner or occupier and accredited representatives of certain Official bodies (see Protection of Birds Act, 1954).

As a result of these legal constraints we have a situation in which farmers feel unduly restricted both in respect of the species they may kill and the methods they may employ to achieve control. At the same time, they are aware that the restrictions have been incorporated into the law by Parliament on the recommendations of bird protectionists and conservationists. This clash of interests and opinion has undoubtedly made a lot of farmers less enthusiastic about wildlife and environmental conservation than they would otherwise be. This is a great pity because the conservation of wildlife needs all the help and goodwill which agriculture can afford to give. The trouble is that, on the one hand, farmers believe that the law should be more liberal in relation to pest control, whilst on the other, the prevailing opinion among bird-lovers tends to favour the imposition of even further restrictions. This problem will not go away, and if we do not do something soon to resolve it, it will get worse. Having spent one-half of my working life as an agricultural scientist and advisory officer and the other as a professional conservationist, I understand both points of view. I am firmly dedicated to the cause of wildlife and environmental conservation, but I hope that this dedication will never distort my perspective nor lead me away from the path of scientific objectivity in considering conservation in the context of all other related environmental problems. I apologise if this sounds pompous, but I think that the best and possibly the only way out of some of our present difficulties is for all of us consciously to exercise more scientific objectivity and common sense in our arguments about pest control than has been apparent in recent years. There seems to me to be a growing trend of conviction that if you want something, you must exaggerate your case out of all proportion to the truth to achieve it. This may succeed occasionally, but the inevitable result is that your opponents will sooner or later adopt the same tactics, so you may end up with stalemate, distrust and reduced ability to communicate. I shall return to this theme presently, but I am sure that exaggeration, or the suspicion of exaggeration, afflicts both sides of many arguments on the issues of bird protection versus pest control.

Council of Europe Directive on the Conservation of Birds

Having touched upon our national legislation, I should like now to say a few words about the implications of the EEC Directive. After over two years of meetings and arguments in Brussels the final text of this document was agreed early in 1979. We are now required to ensure that our own national legislation on bird protection and conservation is brought into conformity with the Directive within the next two years. To meet the EEC deadline we shall have to amend our Protection of Birds Act rather quickly and the Government will have to provide time and facilities to see an amending Bill through Parliament. Since the Act needs amending in various other respects, quite apart from changes made necessary by the EEC Directive, and since the opportunity to amend legislation does not occur very often, it is obvious that those who wish to change the present law will do their utmost to see that their own proposals are incorporated in the new Bill. Interested parties should already be preparing their amendment proposals and briefs for Members of Parliament. Time is short. Although I am a conservationist, I must emphasise that I am not in favour of blanket protection for its own sake, or for sentimental, humanitarian or political reasons. Any protection legislation based on such considerations would be very bad law indeed. I mention this here because the original drafts of the EEC Directive contained much unscientific distortion of fact about declining and endangered bird populations, and the whole text was biased by the subjective views of the bird protectionists who wrote it. I felt strongly at the time that, although the Commission were right to have their legislation on bird protection drafted by ornithologists, they should have chosen people who also had a broad knowledge of ecology and population dynamics and an understanding of the many other conflicting interests in the environment which have to be taken into account – not to mention the requirements of the legislation itself.

Unfortunately the Commission's officials seemed unaware of the subjective imbalance of the document and, during the early months of discussions, they steadfastly defended the draft against what they regarded as unnecessary amendment. As a result, the Working Group of Experts from Member Countries, whose job it was to consider and to reshape the draft Directive into an acceptable form, spent the greater part of two years arguing and bargaining before reaching ultimate agreement. I was a member of the British delegation to this Group, and we, like the representatives from each other country, naturally felt that our own national bird protection legislation was the best and nearest to what the EEC really required. As these national laws and attitudes are all quite different you will appreciate the difficulty of reaching compromise and agreement whilst at the same time preserving as much as possible of what each felt was good and necessary in their own legislation. We had also to bear in mind that the Directive would ultimately override national legislation on bird protection and we wanted to ensure compatibility. Inevitably some sacrifices had to be made but I think that we achieved our objectives and, as a result, I believe that few changes in our current laws will be necessary in order to conform with EEC requirements. This is just another way of saying that we are already doing most of what the Directive requires of us.

There is no time to deal in detail with the provisions of the Directive, but what has emerged is a comprehensive system of protection basically similar to our own. It begins with the statement that all naturally-occurring wild birds are protected, with special protection for those listed in Annex I. The sporting quarry species listed in Annex II pt 1 (24 species) may be hunted throughout the Community and those in Annex II pt 2 (47 species) may be hunted only in specified Member States. Annex III includes a list of seven game species which may be bought and sold for food. Annex IV (a) is a list of prohibited methods of killing or taking birds, and Annex IV (b) limits the use of motor boats and other vehicles for hunting or in pursuit of quarry.

You will note, possibly with dismay, that pest species are not scheduled and not even mentioned as such. Thus, by implication, they remain protected. We fought very hard over a long period to retain a pest annex, especially as one was included in early drafts, listing species which might be hunted or controlled in the interests of agriculture, forestry, fisheries, public health and safety, including air safety, and so on. We were persuaded to accept the majority opinion that a pest annex was not appropriate at Community level but agreed to its omission only on condition that we were assured of the right to derogate, in respect of necessary pest control, from the provisions of Articles 5, 7 and 8 which concern the killing of protected species, killing birds in the breeding season and the use of otherwise illegal methods, respectively, and that Article 9 would be amended accordingly. These conditions were agreed; so the final text of the Directive leaves our pest control arrangements materially unchanged. However, I should point out that Member States are encouraged to carry protection further than required by the Directive, so that our own Government could still curtail the current provisions for pest control. As mentioned, we were dissuaded from pressing what seemed to us a perfectly reasonable issue. We saw much merit in listing the birds which may be controlled where they are too numerous and are causing damage or hazard, as certain species habitually do. Our opponents argued that if birds are "harmful," it is through no fault of their own, so one should not speak of "harmful" birds, or, worse still of "pests"; furthermore, killing birds, except for food, is undesirable and barbarous and it was felt that the Directive should not upset public opinion by including direct reference to pests and their control.

Discussion

Pest problems are man-made, in that man's environmental management, including crop growing and waste-disposal, permits and encourages individual species to multiply to population levels which could never be reached in natural situations. Pollution of the environment, by definition, is a man-made problem, and nobody would argue that man should not be obliged to put it right; weeds in garden and field crops are man-made problems which the grower is well advised to keep in check; the same can be said of rats and rabbits, so why should it not apply equally to the half-dozen or so bird species which have taken advantage of man, through his own fault, and become considerable pests? In my view, man's obligation to rectify this kind of ecological disaster is quite as great as his obligation to control pollution. And, if killing is the only way, then so be it. There need be no fear that any known pest species could possibly be exterminated by using any of the existing and acceptable methods of control.

I do not think that, these days, any reasonable person would argue against the general principle that we should conserve and protect birds and other wildlife and do our best to avoid altering the environment in ways which will reduce their numbers and diversity. However, animal populations are ultimately limited by the carrying capacity of the environment. Since this is true for every species, it is reasonable to assume that, in the absence of other checks, each will increase to the limits imposed primarily by its food resources. Apart from food, secondary environmental conditions, such as the amount of suitable habitat available for successful breeding, and a range of mortality factors, affect the population size for each species.

Even if we established a system under which all birds were fully protected, enabling each species to reach its optimum population size within the constraints already mentioned, it does not follow that the relative abundance or scarcity of most species in Britain would alter significantly, except perhaps for a few pest species. The quantity and carrying capacity of habitats in the British countryside are still the ultimate factors limiting bird populations, and it is unlikely that these can be appreciably improved. Indeed, in view of the economic situation and the still-rising human population, the pressures on every exploitable feature of our environment to provide more food, timber, water, minerals and jobs, plus the indefinitely extended sterilisation of large areas for housing, roads and other permanent developments, make inevitable a continued shrinkage of habitats suitable for birds and other wildlife. In these circumstances, the conservationist has no hope of seeing a future British countryside teeming with richer and more diverse associations of wild plants and animals than we have today. At best, he may hope to reduce the rate at which species are unavoidably lost by establishing nature reserves; to persuade Government and people generally to eliminate unnecessary loss of habitats; to promote legislation and to educate the general public to support wildlife conservation and protection *where necessary*. While I do not think that those needing to control pests will be unduly restricted at the moment by the EEC Directive, I must say that, over and over again in the discussions and debates in Brussels, I was surprised and dismayed by the utterly irrational determination of some delegations to achieve blanket protection at all costs, and to minimise or to delete from the Directive all concessions to sport and pest control. Britain was not alone in resisting this pressure, but you can judge its strength by the fact that we failed to retain a schedule listing bird pests and had to settle for the right to derogate.

The plain fact is that in Europe the interest shown by the general public in birds, and in bird protection, has increased enormously in recent years and we have reached a stage where unreasoning bird-lovers vastly outnumber informed ornithologists. On the face of it, this increase is not a bad thing, as we have long needed a bigger lobby of public opinion to help draw attention to the urgent need for environmental conservation. But the protectionist tail is now beginning to wag the dog and it is possible that, to maintain his nominal leadership, the dog goes where the tail directs, even though he may be well aware that it is ultimately the wrong road, the road which leads only to diminished respect for the responsible conservation bodies and to the loss of their scientific credibility. So, although the EEC influence on bird protection and pest control is not going to oppress us yet, I think that we must remain watchful and be prepared to resist any further legal restriction that is proposed on any grounds other than objectively demonstrated necessity.

Discussion

Thompson: I am sure that I express the feelings of many present in saying how glad I am to see a conference of this kind, bringing together both crop and bird protectors. Norman Moore's reference to the dialogue that has been going on between the Ministry of Agriculture, Fisheries & Food and the Nature Conservancy (now the Nature Conservancy Council) was very welcome: it has been in existence for nearly three decades. Over the past 10 years, in particular, MAFF has been leaning more and more towards an interest in conservation. This has been given an extra impetus by the recommendations of the Srutt Committee Report of May 1978, to which the Minister of Agriculture gave support in principle some two weeks ago. The recommendations to which I refer are that the Agricultural Development and Advisory Service (ADAS), about one-third of the total MAFF staff of some 15 000, should pay increasing regard to the conservation of landscape and wildlife. Although ADAS has a prime responsibility to farmers and agriculture, it is essential to find areas of compromise between agriculture and wildlife and avoid antagonism between bird protectionists and farmers.

Cuthbert: I should like to ask Dr. Boyd how the damage assessments are made in order to arrive at an agreed rate of compensation and whether both the Province administration and the farmers are satisfied with the system.

Boyd: The work is done by a small number of Provincial Assessors who were originally engaged in assessing damage due to hail and their functions have since been extended to look at damage alleged to be caused by all kinds of wildlife in addition to migratory birds. As far as I can make out there is no system of standardising the efficacy of the process, it depends largely on the relatively slow turnover of the Assessors. It is a subjective system but seems to have achieved fairly consistent results. The Province, being responsible for the system, sees no flaw in it and farmers, having learnt to live with it, seem fairly happy too. It is only outsiders who are suspicious.

Wright: In our efforts to assess damage, made over many years, we have always been very conscious of the fact that, from the farmers' point of view, damage that results in a delay in harvesting date is just as important as damage that reduces the actual yield, since missing the market can grossly affect the value of the product.

Armstrong: I agree that assessment of damage and crop loss is difficult. In the case of pink-footed geese and carrots, for instance, it can mean total loss of crop in a matter of days. In Lincolnshire we have had the problem of bullfinches on plums, which in my own case has more or less wrecked plum growing, and this bud damage is now spreading to apples and pears. The other problem we are getting is house-sparrows on black currants; this happens well beyond bud break when there is clearly defined growth. One day the plantation is in fairly good order and the next the whole of the bud population is on the ground; this means virtually a complete loss of crop and I suggest that people who grow black currants are losing a very high proportion of income. Recently there has been less woodpigeon damage but if we get hard weather it may return and this is very difficult to assess. We seem to get a different problem every year.

Wright: I think there is no doubt that we are constantly seeing ecological change, much of it brought about by changes in farming practice, and a shift in emphasis of bird problems as a result. This is well illustrated by the rook, which was considered to be of considerable agricultural importance up to about 1950 but with the changes in agricultural practice introduced after that date its importance declined. Now, with the introduction of maize, and other trends in cereal growing, rooks are again becoming important.

Mount: Is there any evidence that birds change their habits because one or two in a flock develop new tastes. I have been farming a good many years and remember when bullfinches used to go for plums and never worry anything else very much, yet this year they have gone for apples and pears rather than plums. In the 60's we had an influx of redpolls which started going for the pear buds: is this a habit that species can pick-up from a few individuals and so change their source of food?

Flegg: Broadly speaking, I think behavioural trends must spread that way. Blue tits on milk bottle tops is an example that springs to mind and, more recently, magpies in Manchester learning to open the polystyrene egg boxes carried on milk-floats. It takes time for such habits to develop but locally the spread can be quite quick so if you are a fruit farmer it's bad luck to be faced with such a position. Bud-eating habits probably follow the fashions in fruit growing and preferences may be determined by what varieties are grown and *vice versa.* The plum Oulin's gage, for instance, is now virtually out of production because bullfinches like the buds so much.

Bransden: I farm virtually in suburbia, where fragmentation of land, by motorways and such like, is providing false conservation areas where, because access is prohibited, pests can breed and forage unmolested.

Wright: This question of access is one that frequently arises in connection with farm pest control, since, for a variety of reasons, it is often inappropriate to attempt to control a pest at a site where it is doing damage. Your problem is similar to that of someone trying to grow fruit adjacent to woodland he does not own; bullfinch damage could be expected in such a situation but trapping and shooting, to be effective, would need to be undertaken in the neighbouring woodland. If ring-necked parakeets ever become a serious pest in Britain I foresee that it will prove virtually impossible to effect control because of the difficulty of gaining access to their habitat – suburban gardens.

Cadbury: Section 4(2)(a) of the Protection of Birds Act, which concerns the killing of protected species (other than those on Schedule I) presents considerable problems. It states that it shall be a defence in law if the species killed can be shown to have caused serious damage. In such instances as bullfinches eating fruit buds and brent geese grazing cereals in spring, the damage can only be assessed when the crop is harvested, some time later, by which time other factors may have affected the crop.

Colling: I do not entirely share the view that this is a problem; if a farmer cannot assess damage himself then he should be able to call in his advisers and if, in the light of their experience, they say damage is being done that should be adequate evidence to allow the farmer to go ahead and control it; if it is not, then the provisions of Section 4(2)(a) are a farce.

Flegg: I think one can only agree with that; but my concern is more with what the future will bring under the EEC regulations. It is going to be very difficult to present a case that will be acceptable, for the reasons that Dr. Cadbury has pointed out. A lot of factors intervene between bullfinch damage to buds and ultimate yield: frost is one, hail is another and blue tit damage is a third, in these circumstances it is going to be extraordinarily difficult to produce information that will stand up to the kind of scrutiny to which it has been suggested it will be subjected.

Dunnet: I support the need for a scientific basis for decisions regarding pests but there are two immediate problems. Firstly, the difficulty, time and cost of doing the necessary research and, secondly, the difficulty of achieving a political decision as a result of scientific argument. Consider the problem of seals, where a scientific case was overwhelmed by the emotional appeal of Greenpeace. If, under the EEC Directive, pest control depends on derogation and this in turn depends on a scientific case being accepted by politicians, then we are in a very difficult situation and should take steps to safeguard our capability to deal appropriately and effectively with pest species of birds. Assessment of damage by loss of yield is perhaps best but we must make provision for preventing damage rather than simply responding to damage after it has occurred.

Smith: This rather points towards agreeing, within the EEC, some economic damage thresholds for all the crops likely to be affected. Dr. Flegg mentioned that pears can lose 50% or more of buds before the yield of fruit is affected and surely we should be trying to get this sort of information for a lot more crops and get agreement within EEC countries as early as possible.

Boyd: We have not succeeded in reaching agreement in Canada on any scale that can be applied, our assessments still represent a collection of subjective judgements and I think the chances of getting agreement on an objective scale are very remote indeed.

Feare: In the UK we tend to be parochial in our consideration of the number of birds we should have. Most species found here are a western extension of birds with a huge Eurasian range. In the case of those species represented in the UK by less than 100 000 pairs, even if we had the capability of reducing bird numbers, and for the most part we don't have this capability, in the long term this would be unlikely to affect our local populations.

Moore: I agree about the dangers of taking a parochial view but we do have to consider social and scientific benefits of maintaining a good spread of species. For instance, the small numbers of avocets in this country may not be important in terms of world population, but they are important from the conservation point of view, and not least because they are there for people to see. Not so very long ago the Scottish population of golden eagles was an insignificant part of the total European population, but now it is a large part because our birds have been well protected and other golden eagle populations have declined very rapidly.

Smith: Will the introduction of EEC regulations mean that certain species will increase in continental Europe and intensify our problems in the UK?

Wright: The species that you have in mind, the pests, are most unlikely to show any significant changes in populations as a result of man's activities.

Moore: How much is this (the killing of birds to protect crops) a conservation problem? Surely we are dealing with very common species which must allow a good deal of latitude without doing serious harm – or am I wrong about this?

Colling: The main impact of shooting in Europe has been on the rarer species, which have been regularly shot in France and Italy for a very long time. Recent changes in the environment, and other effects that we don't fully understand, have reduced them still further, judging by the records of the number ringed. What we want is a return to what used to be normality; beyond that I don't think it will make any difference.

Wright: Ladies and Gentlemen, lunch is at 1 o'clock and I must reluctantly draw this discussion to a close. It has been something of a warming-up session in which we have considered some of the issues that are peripheral to the central problem of bird damage to agricultural crops. After lunch we shall look more closely at one particular problem – the starling – and tomorrow we shall go on to consider possible ways of tackling these problems. For the moment, however, I should like to thank our four speakers for getting our meeting off to an excellent start, and I invite you to show your appreciation in the usual way.

STARLINGS AS AGRICULTURAL PESTS

Introduction

G. M. Dunnet
The University
Aberdeen

The papers in this session clearly show that the starling can be a serious pest to man's crops in various parts of its European range, and also that it is a bird which lends itself to the detailed study of some basic biological problems. The authors and the audience consist of a mix of people who are concerned with practical problems arising from the unwelcome attention of starlings to man's farming and horticultural activities and others who are primarily concerned with attempting to understand the ecology and behaviour of the bird itself. We all recognise that the very questions that we must ask and, hopefully, answer, require the kind of understanding that can be achieved only by detailed scientific investigation of the starling itself, and the ways it has adapted to rural agricultural environment, as well as a realistic assessment of the costs of starling damage and its control. The ecological problems which any pest species poses are extremely complex, and it is frequently the case that farmers and their advisers have to take action before a full understanding of the situation can be achieved. This can readily lead to such action being built into standard practice without a firm scientific basis, and to the belief that prolonged and painstaking ecological investigation may be unnecessary or even irrelevant. This meeting is an ideal opportunity for scientists to demonstrate that they have an essential role to play in the context of practical and economic problems, and for those who are concerned directly with these problems to identify areas where further study may be necessary.

This afternoon's set of papers highlight the close relationship between practice and theory. Dr. Feare illustrates some of the economic problems that starlings cause in this country and points out the dilemma of the farmer who has to decide whether to apply one set of solutions or another without the full information being available to him or accurate predictions being possible. A major difficulty is to assess what the damage really is; should a farmer wait until it happens in order to measure it or should he assume that it will be severe and attempt to prevent it. This consideration leads on to M. Tahon's paper which describes a drastic method of reducing starling numbers in order to prevent or reduce damage. Even methods as dramatic as he describes rarely have a lasting effect. Professor Verheyen goes on to consider one aspect of the population dynamics of starlings. While he concentrates on breeding biology, equally important is the whole question of mortality and mobility of starling populations. A full understanding of population dynamics would enable us to assess whether the kind of mortality effected by various control measures substitutes for natural mortality or is additive to it. Only if it is the latter, is it likely to be worth carrying out, although in some local short-term situations killing may be justified. When we try to understand the way that wild animals adapt to man-made environments the kind of study that Drs. Drent and Tinbergen present is clearly of the greatest importance. Such work shows the kinds of factors that influence the selection of food by birds. Much theoretical work has recently been done on the foraging efficiency of animals in relation to their energy needs, and this is of the greatest importance in determining the choice which birds make from the range of foods available to them.

I think that the blend of science and practical problems that we have in our programme this afternoon is going to lead to a valuable discussion. I hope it will emphasise the need for communication and collaboration among all concerned to ensure that the scientists can undertake relevant problems leading to the understanding of the practical difficulties in reducing damage to crops by starlings and other pest species.

The economics of starling damage

C. J. Feare
Ministry of Agriculture, Fisheries and Food
Worplesdon Laboratory

Summary

*In Britain starlings (*Sturnus vulgaris*) are responsible for three main kinds of agricultural damage. Their consumption of cherries, estimated at 14–21% of a crop in 1975, results partly from an inability of farmers to regulate the number of pickers according to the amount of ripening fruit. Damage to germinating winter cereals occurs mainly close to roosts when large numbers of starlings may feed briefly on fields in pre-roost assemblies. Although plant densities may be considerably reduced, recent work has failed to show a significant effect on crop yield.*

Damage at feedlots has resulted from the advent of intensive rearing of cattle in enclosed yards. With the food presented to the cattle being freely available to starlings, up to 12% of the barley constituent of the feed may be taken during the winter. Calves housed in enclosures which exclude starlings have been shown to grow faster than animals that have to compete with birds. Although starlings are known to carry, or be capable of carrying, a range of pathogens, good evidence that the birds are involved in disease transmission is lacking.

Since starlings are closely adapted to, and extremely successful in, the agricultural habitats created by man, techniques of damage prevention involving scaring or lethal control are unlikely to succeed without a comprehensive knowledge of the birds' food requirements and population dynamics. It may be more realistic to regard starlings as an environmental factor with which farmers have to cope, and to plan new agricultural developments accordingly.

Introduction

The genus *Sturnus* consists of 16 species of starling characterised by similarities of morphology, behaviour and habitat (Amadon 1943). Their gregariousness, omnivorous diet and preference for open woodland, which man has created in many areas through cultivation, has led several species to become pests. The grey starling (*S. cineraceus*) is a pest of rice in Japan (Kuroda 1973); the Brahminy starling (*S. pagodarum*) takes fruit in Sri Lanka (Henry 1971); the pied mynah (*S. contra*) damages crops (unspecified) in India (Whistler 1963), the rose-coloured starling (*S. roseus*) eats cereals and fruit (Semya 1976, Shi-Chun *et al.* 1975, Henry 1971) and the spotless starling (*S. unicolor*) eats grapes and other soft fruit in Iberia and North Africa. All of these species are, in comparison with the Eurpoean starling (*S. vulgaris*), restricted in their geographical distribution, and although the damage they do to man's crops is sometimes said to be serious, there appear to be no estimates of crop losses in terms of weight of food or loss of money.

The geographical range of the European starling (referred to hereafter simply as "starling") extends over much of western Asia and Europe, breeding as far north as the Arctic Circle and wintering south to North Africa and northern India (Dement'ev & Gladkov 1970, Witherby *et al.* 1940). Within this geographical range starlings encounter a wide variety of crops that they are able to utilise and their diet is further extended where they have been introduced into other parts of the world e.g., Australia, New Zealand, South Africa and North America. Economically valuable foods that they eat include cherries (Brown 1974, Guarino *et al.* 1974, Tahon 1978), olives (Steinbacher 1960, Etchécopar & Hüe 1967), grapes (Henderson 1927, Semya 1976), plums, strawberries and figs (Palmer 1973), blueberries (Mott & Stone 1973), sorghum (Royall 1975), winter-sown wheat and barley (Collinge 1924–27, Stickley *et al.* 1976, Boyce 1979), and cattle food (Besser *et al.* 1968, Feare 1975, Feare & Swannack 1978). In addition to their consumption of food, starlings also pose problems through roosting in plantations and in cities (Elgy 1972, Potts 1967) and it is frequently suspected that they may be involved in the transmission of diseases (D'Alessio *et al.* 1965), although good evidence is lacking (Murton 1964, Snow 1968). They are sometimes regarded as a nuisance when nesting in roof-spaces of houses and inside aircraft hangars – here nesting material and corrosive droppings can damage aircraft and starlings have even nested inside aircraft (Bridgman 1962). Starlings can also be annoying to gardeners by pecking leaves and flowers in order to line their nests (Macgillivray 1840). Furthermore, in countries where starlings have been introduced they may compete with the indigenous avifauna for food and, especially, nest sites (Zeleny 1969, Heusmann & Bellville 1978).

However, despite the wide range of activities that can bring the starling into conflict with man, and the enormous geographical range of the species, estimates of the damage it can cause are scarce. In agricultural terms, damage is best defined as the loss of yield or reduction of financial gain that results from the starlings' depredation of a resource of value to man, but Feare (1974) has shown that attacks by birds on a crop do not necessarily reduce yields and further examples of this will be seen below. In this paper, therefore, the term "damage" is used to refer to situations where starling attacks appear to farmers to be sufficiently serious to result in financial loss. Damage results from the interaction between starling biology and farm management practices, and the ecological implications of both are frequently poorly understood. My aims here are to describe and, as far as possible, quantify the losses due to starlings in Britain and to identify the ecological factors that lead to the conflict between bird and man. This is important if we are to attempt to predict the role of starlings in new agricultural developments.

Starling damage in Britain

The British Isles support two populations of starlings: a largely resident British breeding population, estimated at around seven million pairs (Sharrock 1976), and an immigrant population, which stays from November to March, of around 35 million birds from continental Europe (Potts 1967). Any damage during summer is thus caused by the resident population, while winter problems may be due to residents and/or immigrants. Potts (1967) considered that most urban roosts consisted of resident birds while rural roosts contained mainly immigrants which suggests that agricultural starling problems in winter are most likely caused by the latter. A large winter increase in the number of starlings exploiting a cattle feeding area (=feedlot) at Bridgets Experimental Husbandry Farm, near Winchester, suggested that the majority of these birds were indeed immigrants, although some resident birds, an unknown proportion of which roosted in the farm buildings rather than in local communal roosts, also fed at the feedlots.

Most starling damage results from the birds' feeding activities. The three main areas of damage to British agriculture at present concern ripening cherries, germinating cereals and foods presented to cattle in winter. During much of the nineteenth century the starling was locally distributed in Britain, being most abundant on the Western Isles of Scotland

(Macgillivray 1840). After an increase in numbers towards the end of the nineteenth century, Collinge (1919) remarked on a change from a diet consisting largely of invertebrates, wild fruits and seeds to one containing cultivated fruits and cereals. Collinge (1924–27) also recorded damage to germinating cereals with up to 30% of the plants being removed. However, this type of damage appears to have subsequently declined, although precise data are unavailable, only to have re-emerged as a problem during the last five years, a period when rooks (*Corvus frugilegus*) have also resumed exploitation of this food source (Feare 1978). The advent of damage in cattle feeding areas over the last decade has resulted from a change in agricultural practice, cattle now being housed indoors in large numbers and fed intensively during winter. Further changes in systems of food presentation to cattle have intensified starling problems at feedlots during the last three years.

(a) *Damage to cherries*
Although starlings will eat most soft fruits, and occasionally top fruit, their consumption of ripening cherries constitutes a major problem to fruit growers. The area of cherry orchards in Britain has recently dropped from 7 000 ha to 2 400 ha. This decline has resulted from a variety of factors: the large size of mature trees (leading to high costs of picking), diseases, irregular cropping, fruit splitting and bird damage (MAFF 1977a). Production of cherries in Britain falls short of demand yet imports are prohibited between 1 June and 30 September in an attempt to prevent the entry of cherry fruit fly (*Rhagoletis cerasi*); consequently the home grown cherries that are available command a high price. Any loss to starlings is therefore economically significant. In addition to cherries damaged or eaten by birds, losses are also caused by the fouling of otherwise undamaged fruit and, more important, by the necessity to pick the fruit before it is fully ripe and thus before it has attained its maximum weight.

These factors of fouling and early picking are difficult to quantify but average losses, measured in a 16 ha commercial orchard in 1975, of 14% of cherries on the lower branches and 21% of cherries in the tree canopy indicate the significance of potential losses. The trees studied yielded about 70 kg of cherries per tree after being attacked by starlings and the market price of the fruit was around £1.10/kg. Thus, in 1975 losses were costing around £12 per tree in this orchard and at an approximate 108 trees/ha in commercial orchards (D. Pennell personal communication) this amounts to a loss of £1 296/ha, or £20 736 for the entire orchard, had pickers been available to harvest all the fruit (in fact they were not, see Fig. 1 and below). In 1976 cherry yields were much higher but starling damage was generally less severe. Annual differences were also noted by Brown (1974), but in the absence of detailed information on the availability of alternative foods during each cherry season it is not, at present, possible to explain annual differences in the extent of damage. The damage figures given above for Britain are comparable with those obtained by Brown (1974), Stone (1973) and Tahon (1973) for Canada, United States and Belgium respectively, but they hide the fact that while some trees were left untouched, others were completely stripped by starlings before pickers were able to harvest the crop. Although the birds' preference for certain cultivars (e.g., Van, Merton Glory) plays a part in this, the availability of labour for picking is perhaps the most significant factor. Fig. 1 shows, diagramatically, the availability of ripe cherries, the abundance of starlings and the number of cherry pickers in a commercial British orchard.

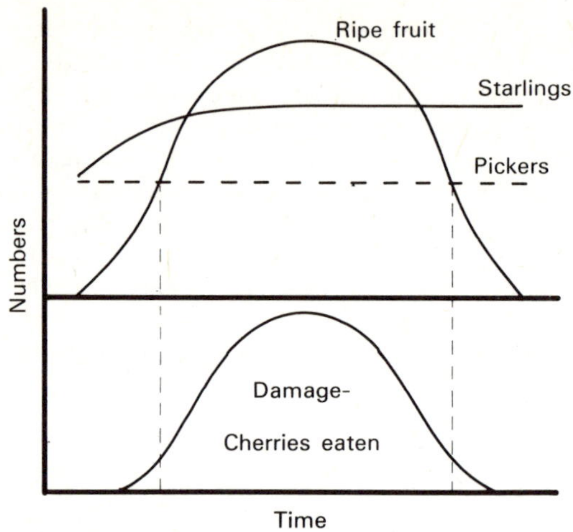

Figure 1. Schematic representation of the quantity of ripe fruit in a cherry orchard in relation to the number of cherry-pickers, and to the damage by Starlings.

The fruits of different varieties ripen in sequence, with most ripe fruit appearing in July, but while the number of starlings increased early in the season, the number of pickers remained constant throughout. While there were few ripe cherries, i.e., early and late, the pickers were able to harvest most of them even though starlings sometimes fed in the same trees where pickers were working. However, at the peak of the season fruit ripened so quickly that pickers were unable to harvest the crop sufficiently rapidly and starlings were able to feed undisturbed in many trees, apart from the control measures attempted by the grower. Thus, in this situation proportionally more damage was done when the "prey" (ripe cherries) was most abundant – an unusual situation in predator-prey relationships. But this resulted from the cropping regime practised by the grower: if the number of pickers could be related to the abundance of ripe fruit the losses to starlings would be considerably reduced. This is obviously difficult with high labour costs.

Fig. 2 shows that most of the starlings feeding in the orchard in 1975 were juveniles, although there was a diurnal cycle with adults, possibly locally roosting birds, predominating in the early morning and late evening. The predominance of juveniles in cherry orchards has been noted elsewhere (Brown 1974, Tahon 1978) and limited data collected in southern England suggest that while adults remain predominantly terrestrial feeders juvenile starlings are mainly arboreal (Table 1). This predisposition to arboreal behaviour of juveniles may result partly from their being less conspicuous in a tree canopy. However, the juveniles utilisation of fruit, in contrast to the adults' apparent preference for grassland invertebrates, could also reflect different nutritional requirements of the age groups and an attempt to reduce intra-specific competition. After the breeding season (in southern England it usually finishes in late May–early June) adults have low protein and fat reserves (Ward 1977) but despite this they begin to moult. Soil invertebrates are scarce in summer (Dunnet 1955, Dunnet & Patterson 1968, Feare *et al.* 1974) and naive juveniles may, by exploiting arboreal resources, avoid competition with more experienced adults for the scarce soil fauna. Juveniles moult somewhat later than

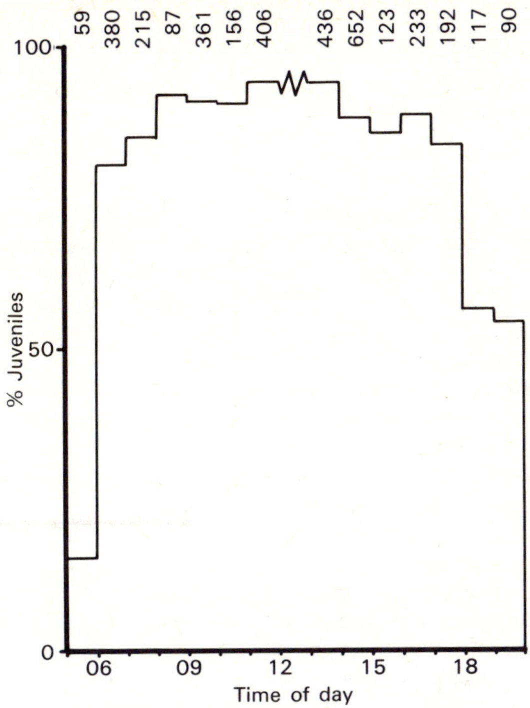

Figure 2. The age composition at different times of day of flocks of Starlings in a commercial cherry orchard. Figures at top are sample sizes.

adults (Svennson 1975) and their main protein requirement may come later (Kendeigh *et al.* 1977) nearer the time when the abundance and availability of soil invertebrates increases. Brown (1974) suggested that the water content of cherries could be important to starlings, especially in times of drought, but this would presumably affect both adults and juveniles.

Starling damage to cherries is thus influenced both by the requirements of the birds and orchard management. Juvenile birds are mainly responsible for damage and they show a propensity for arboreal feeding during the summer. The basis of their preference for

Table 1

The proportions of juvenile Starlings in ground and tree feeding flocks at Bridgets EHF, near Winchester, in July and August 1975 (no soft or top fruit is grown commercially in this area). $X^2 = 607{\cdot}5$ $p \leqslant 0{\cdot}001$

	No. of flocks recorded	No. of adults	No. of juveniles	% juveniles
Ground	26	895	92	9·32
Trees	13	13	164	92·66

Table 2

Counts of Starlings on six winter wheat fields in Lincolnshire

Time of day	Sunrise–0900	0900–1200	1200–1500	1500–sunset
Mean number of Starlings	86	57	107	49 071
SE	46	57	64	27 095
RANGE	0–300	0–400	0–400	0–200 000

certain cultivars is not known. Losses could be reduced if growers were able to regulate the size of their labour force according to the number of ripe trees to be picked. These factors are all relevant when considering the most appropriate control measures.

(b) *Damage to germinating winter cereals*

Starlings usually begin eating seeds of autumn-sown wheat and barley shortly after the coleoptiles have emerged above ground level, but fields are occasionally attacked before emergence. Although some seeds are obtained directly from the soil, leaving the plant apparently intact, others are obtained by uprooting the plant before removing the grain: in this case the plant remains are discarded on the soil surface.

Most attacks occur in November–December, and it is late-sown fields, especially those close to winter roosts, that suffer; the critical factor is probably the autumn arrival of continental immigrants and the formation of their large rural roosts, rather than late sowing *per se*. Damage close to winter roosts is done mainly in the evenings when birds congregate in large pre-roost assemblies prior to entering the roost (Table 2). These flocks, containing up to 200 000 birds, may spend only a short time on the fields and considerable damage can be done in a few minutes. Wadsworth (unpublished MAFF report) recorded probing rates of $3{\cdot}8 \pm 0{\cdot}2$/min, although not all were followed by a swallowing movement that indicated food had been obtained. Fields close to the roost were not used regularly every day, but in a field of winter barley, sown in October, adjacent to a sewage treatment works, small flocks (up to 100 birds) were observed feeding all day; this continued for several days and led to extensive plant loss.

Although proximity to a large winter roost appears important in determining the probability of a field being attacked, the factor that seems to have been responsible for the recent increase in this kind of damage is a tendency for farmers to sow their seed less deep. Shallow planting allows more rapid plant establishment and also assists in combating attacks by wheat-bulb fly (*Delia coarctata*), which are more prevalent in late-sown crops (MAFF 1977b). Boyce (1979) has demonstrated a relationship between depth of sowing and the probability of seed removal by starlings, with seed deeper than 50 mm not being removed. Starlings' bills average 30–35 mm, but the abrasion of facial feathers during the breeding season suggests that they will probe deeper than this. In areas where both wheat-bulb fly and starling damage might be predicted, the farmer must assess the risks and sow accordingly.

At present there are few data on the effects of starling damage on yield of cereal crops. Wadsworth found reductions in wheat plant density of 23–37% but found no significant differences in number of ears, 1 000 grain weights or total yield between protected and unprotected plots. In the field of winter barley I recorded an overall reduction in plant density of 29%, although in localised patches density was reduced by 90%. However, as a result of increased tillering there were more ears per plant in damaged plots than in the undamaged samples. This compensatory growth resulted in there being no significant differences between the yields of damaged and undamaged areas. As in Wadsworth's study, starling damage appeared not to result in a reduced yield but in both of these studies the fields had been sown at a rate 5–10% above normal: the wheat to compensate for anticipated starling damage and the barley to compensate for late sowing of the crop. Studies are obviously required of the effects of starlings on the yields of crops sown at lower densities, since Feare (1974) found that rooks considerably reduced yields of some spring-sown barley and oats in north-east Scotland. There have been some MAFF reports of fields requiring re-drilling after starling attack, although it is not certain that the damage was due entirely to this species. Where fields are over-sown in an attempt to offset losses there is a possibility that the high grain densities actually attract birds to feed there: this also needs investigation.

For autumn-sown cereals, therefore, it is not at present possible to quantify starling damage, but where seed rates are increased, as in Wadsworth's study, from around 180 kg/ha to over 200 kg/ha in order to compensate for expected bird damage, the cost would be about £4/ha.

(c) *Damage associated with livestock feeding areas*
Potential damage at animal feeding areas falls within two categories: (i) the removal of animal food, and (ii) problems associated with fouling and transmission of disease. While the removal of food is fairly readily quantified (but has rarely been done), problems related to fouling and disease transmission are difficult to measure or even demonstrate satisfactorily.

(i) Removal of animal food
Most estimates of losses of animal foods to starlings concern cattle food, but Murton & Westwood (1976) quoted a loss of £33 000 per annum of food at a large poultry farm in eastern England.

In the USA, Besser *et al.* (1968) estimated that starlings took $84 of cattle food per 1 000 birds in one winter, while in England Wright (1973) demonstrated that calves in pens protected from starlings showed higher growth rates and better food conversion than those in unprotected pens, leading to an increase in profit margin of £3 per head. Food loss was not measured directly.

Feare & Swannack (1978) estimated that food eaten by starlings from a calf-rearing unit at Bridgets Experimental Husbandry Farm over three winters amounted to 6–12% (£125–£222) of the food presented to the calves. Since then two further estimates have been obtained with 4% (£70) and negligible damage occuring respectively. During these studies the numbers of starlings on the farm, their feeding sites, the timing of utilisation of the calf-rearing areas and the sex/age composition of the birds involved have been monitored. The number of birds using the farm each year and their time of occurrence (Fig. 3) has varied. In the winters of 1974–75 and 1977–78 the maximum number on the farm was reached in November, followed by a decline until March, when all the immigrants left. In 1975–76 numbers remained high after the November peak, while in 1976–77 the maximum population was attained in January and in 1978–79, after a "normal" November arrival, numbers quickly dropped to the level expected in March, indicating that the immigrant population did not stay on the farm.

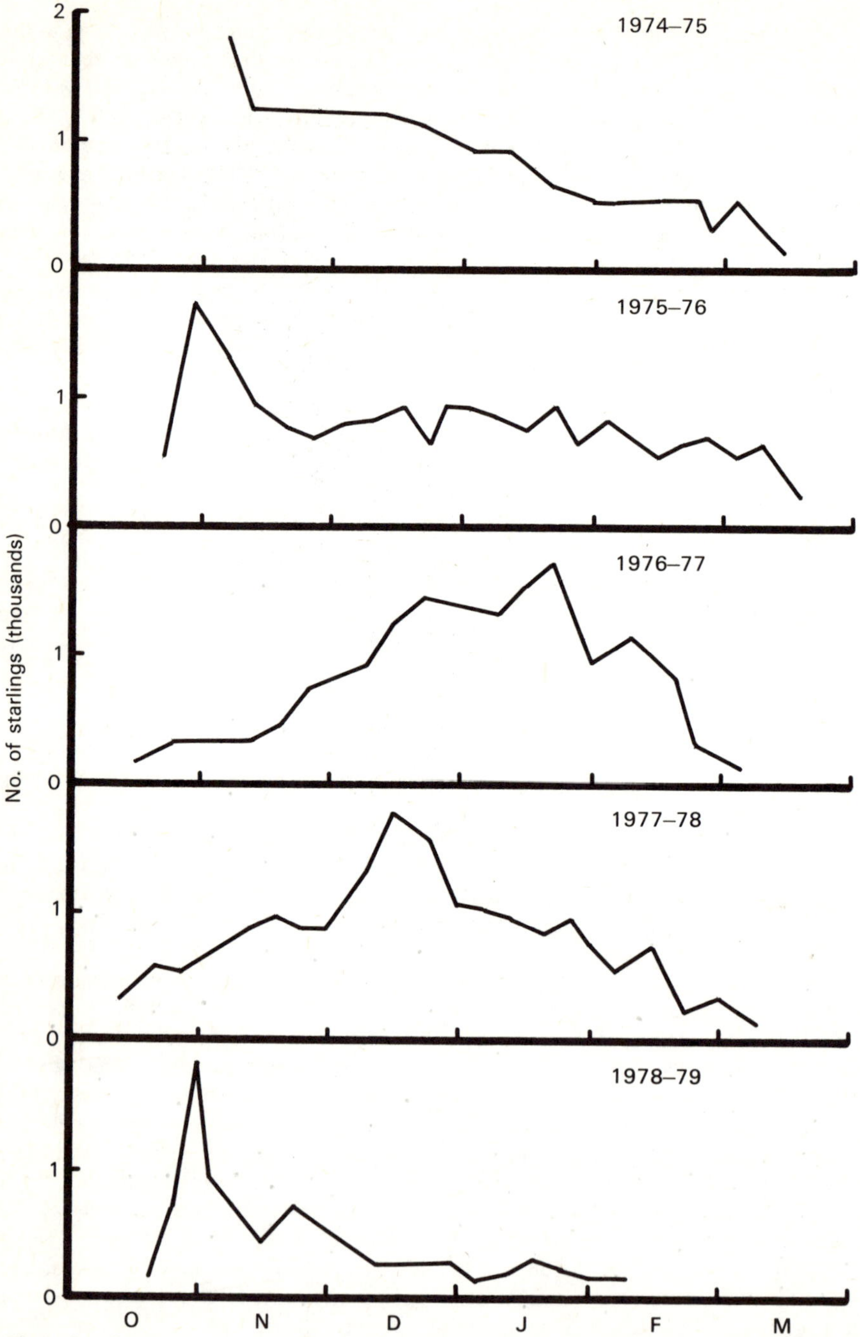

Figure 3. Seasonal changes in the number of Starlings at Bridgets Experimental Husbandry Farm.

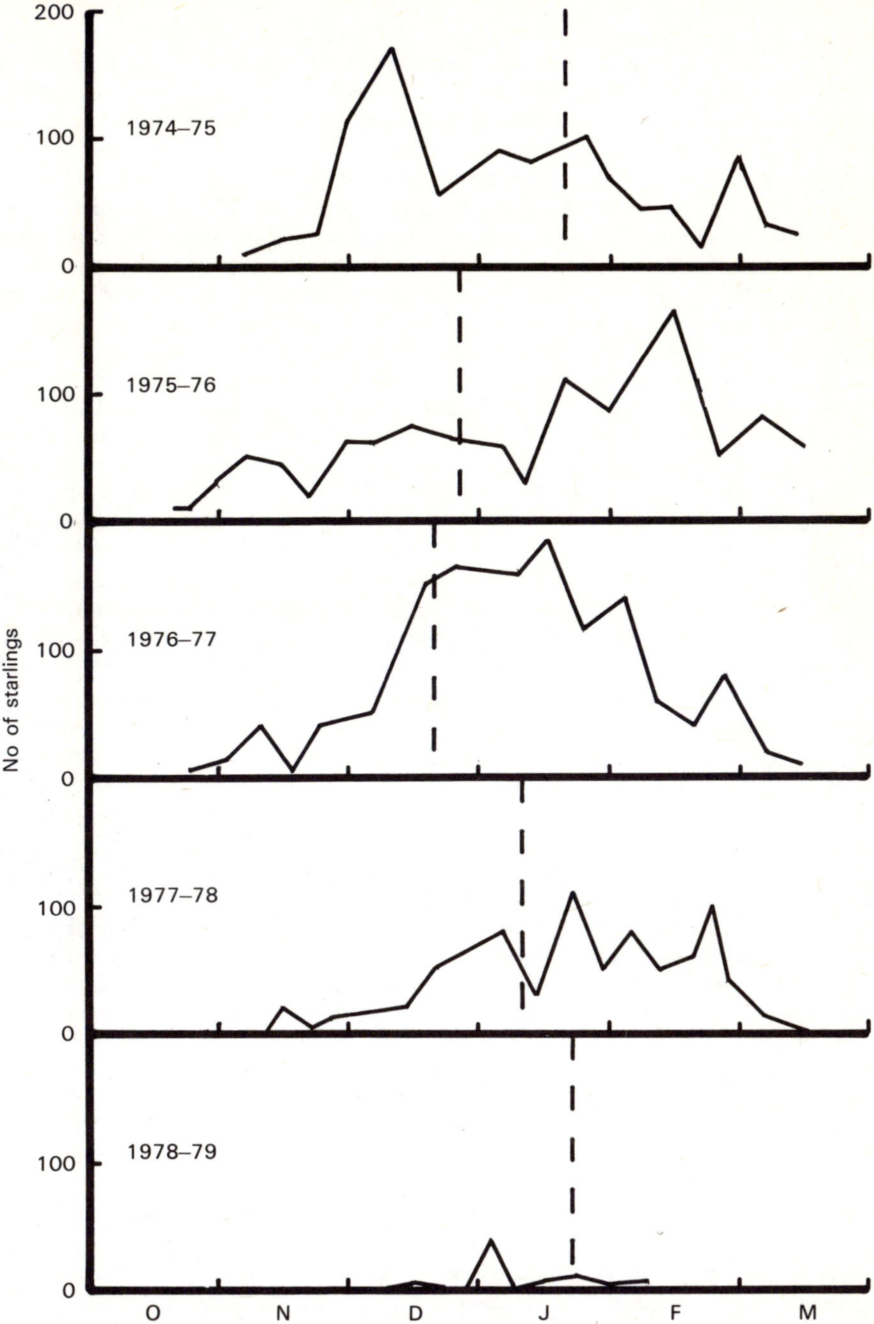

Figure 4. Seasonal changes in the number of Starlings feeding in the calf unit at Bridgets Experimental Husbandry Farm. Vertical line denotes change of diet.

The number of birds that fed in the calf unit did not parallel the total number of birds counted on the farm (Fig. 4). For example, in 1974–75 the large numbers present in November fed mainly on maize stubble and only began using the calf unit when the field was ploughed; peak numbers were recorded in the feedlot in early December (Feare 1975). In 1975–76 and 1977–78, most was taken from the feedlot in January–February, while the situation in 1976–77 was intermediate. In this calf unit the timing of the birds' utilisation of the calf food to a certain extent determines the value of the food that is lost. In December or January the calves' diet was changed from one consisting solely of starter pellets to one containing a 3:1 mixture of crushed barley:concentrate pellets from which starlings selected the barley (Feare 1975, Feare & Swannack 1978). Barley is less expensive than either the starter or concentrate pellets so that food loss after the change of diet is less costly than that occuring earlier.

It is impossible to predict the number or seasonal pattern of occurrence of starlings on the farm or when they will feed in the feedlot. In January 1979, when few birds were present on the farm, a period of deep snow led to an increase in the numbers in the feedlot, but in other years really severe weather was not experienced. Data on the distribution on the farm of feeding birds during the five winters of this study indicate that $68{\cdot}9 \pm 1{\cdot}7\%$ (n=223 transects) fed in grassland. There were no statistically significant seasonal or diurnal variations in this proportion, except that on mornings with 5 cm soil temperatures below 0°C only $2{\cdot}2 \pm 0{\cdot}8\%$ (n=10) were feeding in grass fields, most birds using other farmyard food sources (slurry, dung heaps, food presented to dairy cattle etc).

However, marking studies (Feare 1978, in press) have shown that starlings feeding in the calf unit represent a biased sample of the population, consisting predominantly of males. I interpreted this as signifying that males were dominant, and have a competitive advantage over females in the crowded conditions found at the feeding troughs. This has been demonstrated experimentally in captive flocks of starlings (Van der Mueren 1977, Feare & Inglis 1979). Although samples of birds from grassland sites have not yet been obtained, a sample from a dairy cow feeding area, with almost unlimited space, contained over 50% females. The selection of what are possibly prime (i.e., high energy intake) feeding areas by a relatively small section of the starling population has implications for both potential disease transmission and attempts to prevent damage.

The removal of food from farms where calves are reared intensively, as at Bridgets EHF, has presented problems for over a decade, but during the last five years a new development in the feeding of dairy cows in Britain has led to unforeseen starling problems which are potentially even more costly. The "complete-diet" feeding of dairy cows involves the presentation to cattle, restricted in yards and/or buildings, of "an intimate mixture of processed ingredients presented in a form which precludes selection and which is designed to be the sole source of food" (MAFF 1977, Rickaby 1978). Two investigations by the Agricultural Development and Advisory Service (ADAS) (MAFF 1977, 1978) have highlighted the damage caused by starlings, but studies to assess this damage are in their infancy. However, preliminary results from a commercial farm near Huntingdon have shown that losses can be high. Samples of starlings shot at the feeding trough were predominantly males (88%, n=60) and their gizzards contained fragments of barley with occasional short pieces of grass silage, possibly swallowed incidentally, indicating that crushed barley was selected from the ration. Thus, even though the space available for feeding was much greater than in the calf troughs it appears that there was sufficient competition for males to exclude females. From the number of birds feeding at the trough, their feeding rate and the size of the food particles selected, it was estimated that about 11·9 tonnes (value c.£1 000) of barley had been taken from the 120 tonnes fed to the herd of 140 dairy cows over four months. The effects of this loss of barley from the ration on the cows and their milk yield is not yet known. Some farmers claim that milk yields fall in these circumstances and this is attributed to the loss of a high energy fraction of the diet, to stress caused by the presence of large numbers of birds and to decreased palatability of food when contaminated with bird droppings. Data on cow behaviour are required before these claims can be substantiated. Nevertheless, the loss of barley to starlings is itself a major problem to complete-diet feeding systems. It costs an average of £40 per animal to introduce complete-diet feeding but, as has been pointed out (MAFF1978) the system does not offer overall financial rewards and farmers must take advantage of other benefits, such as convenience and ease of management.

(ii) Fouling and disease

Although fouling can be dangerous where droppings accumulate on pavements or industrial structures, the main complaint around farm buildings is that fouling produces unpleasant working conditions. During the winter guano rapidly accumulates on feed troughs, walls, gates and other fencing within buildings where starlings feed; the hair of the cattle may also become matted with droppings. Possible effects of droppings on food palatability have already been mentioned.

The role that starlings play in the transmission of diseases, as mechanical vectors or through their own infection, is obscure. Some recent trends in husbandry practice, especially the indoor rearing of pigs and poultry, have reduced the potential for transmission of diseases like Newcastle Disease and Duck Virus Hepatitis by birds. Nevertheless, starlings are known to carry pathogens although the incidence of infected birds is usually low. For example, Wilson & Macdonald (1967) reported an incidence of only 0·4% of *Salmonella typhimurium* in a sample of 540 starlings collected in winter, and J. Scudamore (personal communication) failed to find pathogens or antibodies during examination of 42 starlings trapped in the calf feedlot at Bridgets EHF during the 1976–77 winter. Even at sewage farms *Salmonella* infections in birds are rare (Plant 1978) but samples of birds found dead in roosts tend to have a higher incidence of pathogens (Faddoul *et al.* 1968).

Where outbreaks of bacterial diseases have occurred in farm stock the causative organisms have sometimes been found in starlings on the same farm. Bickford *et al.* (1966) isolated *Mycobacterium avium* from starlings at a farm in the United States where pigs had contracted tuberculosis, while in Britain, P. Bloxham (personal communication) has recently found a high proportion of starling samples infected with *Leptospira laura* at a farm where this organism was causing abortion in cattle. However, the relation between the infections in the birds and the stock is unknown.

With virus diseases, especially where birds act only as mechanical disseminators, the role of starlings is even more difficult to determine. Murton (1964) reviewed the many factors that could operate to introduce foot and mouth desease (FMD) into Britain and lead to its spread from farm to farm and concluded that starlings were unlikely to be implicated. Subsequently, Snow (1968) found no correlation between the movement of birds and the spread of FMD in an area close to a starling roost, but he did find a relation between direction of spread and the direction of strong prevailing winds. Pilchard (1965) successfully infected pigs with Transmissible Gastro-Enteritis (TGE) using starling faeces collected after the birds had been fed suspensions of infected pig intestine; however, the evidence that starlings transmit the disease from farm to farm is only circumstantial: contaminated lorries and the movement of pigs or humans may be more serious sources of infection. Similarly, there is little good evidence that wild birds are involved in the spread of influenza viruses to poultry (Easterday 1975), even though Lipkind *et al.* (in press) have isolated Influenza A, serotype Hav 1 Neq 1, from a starling in Israel and Alexander *et al.* (1979) have isolated the same virus from turkeys with respiratory disease and poor egg production. The significance of the discovery by Aspock *et al.* (1972), of antibodies to the arboviruses Tahyna, Yellow Fever and Dengue II in starling blood, is unclear but it does indicate that these birds can be infected by a wide range of viruses.

Starlings appear to be implicated in other veterinary and public health problems: Haslett & Schneider (1978) found antibodies to *Toxoplasma gondii* in starlings in an area where stock reared for human food was found to be seropositive; Dodge (1965) and Tosh *et al.* (1970) have suggested that the contamination of starling and blackbird (Icteridae) faeces with *Histoplasma capsulatum* was responsible for histoplasmosis infections in schoolchildren living near roosts, and Crewe & Owen (1978) have intimated that starlings could be involved in the dissemination of eggs of the human beef tapeworm (*Taenia saginata*). Starlings carry a wide range of helminith parasites (Owen & Pemberton 1962, Cooper & Crites 1976) but most of these are of no agricultural significance. The epizootiology of all of these diseases requires further research before the role of starlings can be ascertained.

Management of starling damage

We know little about the cost of starling damage and even less about the economics of preventing or reducing that damage. In fact, there have been few attempts to assess the effectiveness of the many devices, some of them expensive, currently used by farmers in attempts to resolve problems caused by starlings and other birds. For example, in the commercial cherry orchard studied, the farmer attempted to reduce starling damage by shooting, using sirens on cars, rattles, propane bangers and other noise-making equipment. The cost of this, in terms of cartridges and labour, was estimated at about £2 000 but despite these measures starlings consumed cherries worth over £20 000. At the calf-rearing unit at Bridgets Experimental Husbandry Farm, however, Feare & Swannack (1978) demonstrated that food losses to starlings could be prevented by various means of proofing. The most effective and convenient material, heavy-duty PVC strips, was expensive, but with a life of several years was economically worthwhile. Experiments on the cost-effectiveness of proofing complete-diet dairy units are now in progress. Nevertheless, some agricultural systems that suffer damage are not amenable to proofing and other forms of management – either of the bird population or the farming activity – have to be considered.

Management of starling populations can consist of a reduction of the population or of a change in the feeding dispersion of individuals so that they no longer eat vulnerable crops. When considering possible methods of achieving either of these aims it is important to realise that starlings, like all bird pests, possess attributes that have enabled them to succeed in man-modified habitats. Of the 21 ecologically important variables of bird pests considered by Weins & Johnston (1977), starlings possess 14, of which the major ones, with reference to British problems, are probably communal roosting, flock feeding, food opportunism, high population turnover and migration. Being so closely adapted, by these and other characteristics, to cultivated habitats, any attempt to alter population levels or feeding dispersion, without at the same time altering the habitat, is inevitably extremely difficult. Furthermore, if the habitat, or in the agricultural context the attractive source of food, is not changed the solution to the bird problem is likely to be only temporary.

Population reduction, involving the destruction of large numbers of starlings, is the strategy usually advocated by farmers to reduce damage. However, where such techniques have been applied they have been found to be only partially effective, and temporary (e.g., see Tahon, this volume). Even for a bird as abundant as the red-billed quelea (*Quelea quelea*), local control aimed at the reduction of damage is preferable, on both ecological and economic grounds, to attempted reduction of the entire population (Ward 1973, Dyer & Ward 1977). Tahon (1977) has shown that a comprehensive knowledge of the extent and rate of local movements of starlings is required to assess the efficacy of lethal control in reducing damage to cherries. Current knowledge of the local movements of starlings in winter in the vicinity of cattle feeding areas poses further questions that require answers before lethal control is attempted. For example, since the starlings visiting feedlots appear to represent only a segment of the population, consisting largely of males which show high fidelity to their feeding areas, (Feare in press, Feare & Swannack 1978), adequate reduction of damage might be achieved simply by removing those birds found feeding at the calf unit. If the remainder of the farm population is tied to invertebrate food by nutritional requirements this control may be sufficient, but if the females are excluded by the males through competition for feeding space (Feare & Inglis 1979), once the dominant birds have been removed the subordinates will probably move in to take their place. In this case a much larger part of the local population would have to be killed in order to achieve effective control of damage but the creation of such a void could again lead to immigration from surrounding areas. If immigration was rapid, a lethal control strategy might prove unacceptable on economic, as well as social and environmental grounds, depending on the techniques used.

Killing of starlings in pre-roost assemblies also requires further knowledge of their behaviour. Two kinds of assembly are formed, one consisting of feeding birds and the other of flocks of resting or preening birds. The difference, if any, in the social or nutritional status of birds in each kind of assembly is unknown; nor is it known whether there is extensive interchange between the groups, or whether the same individuals are involved in the same kind of assembly each evening. This information is required before an appropriate control strategy can be formulated: should birds be killed in the feeding pre-roost assemblies (West 1968), or should the entire roosting population be destroyed (Steinbacher 1960, Lefebvre & Seubert 1970)?

The alternative to lethal control of a starling population is to modify the distribution of feeding birds in the agricultural ecosystem so that they no longer feed in places where damage might result. The ability to achieve this modification presupposes that alternative, nutritionally adequate, feeding sites are available to the birds. When damage is being done, alternative resources are frequently not available (Dyer & Ward 1977, Feare 1974, 1978, Feare, Dunnet & Patterson 1974, Newton 1964, Ward 1965), which accounts for the failure of many scaring devices, chemical repellents and "bird-resistant" varieties of crops.

The provision of alternative food resources for birds, as a way of preventing damage, necessitates knowledge of the requirements of different age and sex groups, in terms of both feeding sites and nutrition, both of which vary seasonally. This knowledge might also permit an understanding of the annual variations in the severity of damage, as seen in cherry orchards and in feedlots, possibly permitting the prediction of "bad" and "good" years when control may or may not have to be exercised.

However, in view of the difficulties of controlling starling populations that are ecologically adapted to current farming practices, a more realistic and long-term solution to damage problems could be to seek methods of managing the agricultural habitat to make it less attractive to these birds. The housing of complete-diet dairy units inside bird-proof buildings offers additional advantages apart from protection from starlings (Feare & Swannack 1978). With the complete-diet dairy system, where financial gains are marginal, food losses to birds were predictable and the anticipation of damage would have helped planners to combat the problem at an early stage. A possible alternative to indoor housing follows anecdotal evidence that a change in the constituents of the complete-diet ration, e.g., the substitution of waste grain from breweries for crushed barley, can, in some instances, reduce losses to starlings. Here again, knowledge of the birds' nutritional requirements could play an invaluable part in the formulation of food rations.

The recent resurgence of starling (and corvid) consumption of sown and emerging winter cereals, again apparently following a change in husbandry techniques, might also have been predicted: Weins & Johnston (1977) considered the starling to be "not yet a granivorous bird pest" but that it could easily become one. This example further highlights the requirement for liaison between ornithologists and agricultural advisers.

Although attempts to kill and scare bird pests are traditional in agricultural societies, it is becoming increasingly apparent that many of the techniques fail because they are based on untenable assumptions, e.g., that increasing control effort will inevitably lead to a reduction of damage (Murton & Jones 1973, Dyer & Ward 1977). A more practical alternative may well be to regard a bird pest simply as another environmental factor that must be taken into account when considering modifications to existing husbandry techniques or the introduction of new ones. The ability to view birds in this light presupposes, of course, a detailed knowledge of the relationship between the bird and its agricultural habitat.

References

ALEXANDER, D. J.; ALLAN, W. H.; PARSONS, G. (1979) Characterisation of influenza viruses isolated from turkeys in Great Britain during 1963–1977. *Research in Veterinary Science* 26, 17–20.

AMADON, D. (1943) The genera of Starlings and their relationships. *American Museum Novitates* 1247, 1–16.

ASPOCK, H.; GRAEFE, G.; KUNZ, C.; RADDA, A. (1972) Antikörper gegen Arboviren in Staren *(Sturnus vulgaris* L.) in Österreich. *Zentralblatt fur Bakteriologie* 221, 141–142.

BESSER, J. F.; DE GRAZIO, J. W.; GUARINO, J. L. (1968) Costs of wintering Starlings and Red-winged Blackbirds at feedlots. *Journal of Wildlife Management* 32, 179–180.

BICKFORD, A. A.; ELLIS, G. H.; MOSES, H. E. (1966) Epizootiology of tuberculosis in starlings. *Journal of the American Veterinary Medical Association* 149, 312–318.

BOYCE, D. V. H. (1979) The influence of sowing depth on the removal of grain from winter wheat by Starlings (*Sturnus vulgaris* L.). *Journal of Plant Pathology* 28, 68–71.

BRIDGMAN, C. J. (1962) Birds nesting in aircraft. *British Birds* 55, 461–470.

BROWN, R. G. B. (1974) Bird damage to fruit crops in the Niagara Peninsula. *Candian Wildlife Service Report* 27, 1–57.
BULLOUGH, W. S. (1942) On the external morphology of the British and Continental races of the Starling (*Sturnus vulgaris* L.). *Ibis* 6, 225–239.
COLLINGE, W. E. (1919) The plague of Starlings. *The National Review* April 1919. 252–257.
COLLINGE, W. E. (1924–1927) The food of some British wild birds. York; Published by the Author.
COOPER, C. L.; CRITES, J. L. (1976) Additional check list of the helminths of the starling (*Sturnus vulgaris* L.). *American Midland Naturalist* 95, 191–194.
CREWE, B.; OWEN, R. (1978) 750 000 eggs a day – £750 000 a year. *New Scientist* 80, 344–346.
D'ALESSIO, D. J.; HEEREN, R. H.; HENDRICKS, S. L.; OGILVIE, P.; FURCOLOW, M. L. (1965) A starling roost as a source of urban epidemic histoplasmosis in an area of low incidence. *American Review of Respiratory Diseases* 92, 725–731.
DEMENT'EV, G. P.; GLADKOV, N. A. (1970) Birds of the Soviet Union Volume 5. Jerusalem: Israel Program for Scientific Translations.
DODGE, H. J. (1965) The association of a bird-roosting site with infection of schoolchildren by *Histoplasma capsulatum*. *American Journal of Public Health* 55, 1203–1211.
DUNNET, G. M. (1955) The breeding of the Starling *Sturnus vulgaris* in relation to its food supply. *Ibis* 97, 619–662.
DUNNET, G. M.; PATTERSON, I. J. (1968) The Rook problem in north-east Scotland. In Murton, R. K.; Wright, E. N. (*Eds*) The problems of birds as pests. London; Academic Press. p. 119–139.
DYER, M. I.; WARD, P. (1977) Management of pest situations. In Pinowski, J.; Kendeigh, S. C. (*Eds*) Granivorous birds in ecosystems. Cambridge; University Press. p. 267–300.
EASTERDAY, B. C. (1975) Animal influenza. In Kilbourne, E. D. (*Ed*) Influenza viruses and influenza. New York; Academic Press. p. 449–481.
ELGY, D. (1972). Starling roost dispersal in forests. *Quarterly Journal of Forestry* 66, 224–229.
ETCHÉCOPAR, R. D.; HÜE, F. (1967) The birds of North Africa. Edinburgh; Oliver & Boyd.
FADDOUL, G. P.; FELLOWS, G. W.; BAIRD, J. (1968) Erysipelothrix infection in Starlings. *Avian Diseases* 12, 61–66.
FEARE, C. J. (1974) Ecological studies of the Rook (*Corvus frugilegus* L.) in north-east Scotland. Damage and its control. *Journal of Applied Ecology* 11, 897–914.
FEARE, C. J. (1975) Cost of Starling damage at an intensive animal husbandry unit. *Proceedings of the 8th British Insecticide and Fungicide Conference* 1, 253–260.
FEARE, C. J. (1978) The ecology of damage by Rooks (*Corvus frugilegus*). *Annals of Applied Biology* 88, 329–334.
FEARE, C. J. (1978) A wing tag for Starlings. *European Plant Protection Organisation Publications Series B* 84, 51–54.
FEARE, C. J. (In the Press) Local movements of Starlings in winter. *Proceedings of the XVII International Ornithological Congress, Berlin.*
FEARE, C. J.; DUNNET, G. M.; PATTERSON, I. J. (1974) Ecological studies of the Rook (*Corvus frugilegus* L.) in north-east Scotland. Food intake and feeding behaviour. *Journal of Applied Ecology* 11, 867–896.
FEARE, C. J.; INGLIS, I. R. (1979) The effects of reduction of feeding space on the behaviour of captive Starlings *Sturnus vulgaris*. *Ornis Scandinavica* 10, 42–47.
FEARE, C. J.; SWANNACK, K. P. (1978) Starling damage and its prevention at an open-fronted calf yard. *Animal Production* 26, 259–265.
GUARINO, J. L.; SHAKE, W. F.; SCHAFER, E. W. (1974) Reducing bird damage to ripening cherries with methiocarb. *Journal of Wildlife Management* 38, 338–342.
HASLETT, T. M.; SCHNEIDER, W. J. (1978) Occurence and attempted transmission of *Toxoplasma gondii* in European Starlings (*Sturnus vulgaris*). *Journal of Wildlife Diseases* 14, 173–175.
HENDERSON, J. (1927) The practical value of birds. New York; Macmillan.
HENRY, G. M. (1971) A guide to the birds of Ceylon. Oxford; University Press.

HEUSMANN, H. W.; BELLVILLE, R. (1978) Effects of nest removal on Starling populations. *Willson Bulletin* 90, 287–290.

KENDEIGH, S. C.; DOL'NIK, V. R.; GAVRILOV, V. M. (1977) Avian Energetics. In Pinowski, J; Kendeigh, S. C. (*Eds*) Granivorous birds in ecosystems. Cambridge; University Press. p. 127–204.

KURODA, N. (1973) Fluctuation of winter roosting flock of *Sturnus cineraceus* at Koshigaya and the roost change to Omatsu in summer. *Miscellaneous Reports of the Yamashina Institute of Ornithology* 7, 34–55.

LEFEBVRE, P. W.; SEUBERT, J. L. (1970) Surfactants as Blackbird stressing agents. In Dana, R. H. (*Ed*) *Proceedings of the 4th Vertebrate Pest Conference.* West Sacramento; California Vertebrate Pest Committee. p. 156–161.

LIPKIND, M. A.; WEISMAN, Y.; SHIHMANTER, E.; SHOHMAN, D.; DOUGLAS, A.; SKEHEL, J. J. (In the Press) Characterisation of avian influenza viruses isolated in Israel in 1978 and 1979. *Symposium International Grippes Animales et Humaines.*

MACGILLIVRAY, W. (1840) A manual of British Ornithology. *Part I. The Land Birds.* London; Scott, Webster & Geary.

MAFF (1977a) Sweet cherries – progress towards a revival. *Short-term Leaflet* 196, 1–9.

MAFF (1977b) Wheat Bulb Fly. *Advisory Leaflet* 177, 1–5.

MAFF (1977) Complete diet feeding of dairy cows. London; Her Majesty's Stationery Office.

MAFF (1978) Second year report on complete-diet feeding of dairy cows. London; Her Majesty's Stationery Office.

MOTT, D. F.; STONE, C. P. (1973). Bird damage to blueberries in the United States. *Bureau of Sport Fisheries and Wildlife Report* 172, 1–15.

MURTON, R. K. (1964) Do birds transmit foot-and-mouth disease? *Ibis* 106, 289–298.

MURTON, R. K.; JONES, B. E. (1973) The ecology and economics of damage to Brassicae by Wood-pigeons. *Annals of Applied Biology* 75, 107–122.

MURTON, R. K.; WESTWOOD, N. J.; (1976) Birds as pests. *Applied Biology* 1, 89–181.

NEWTON, I. (1964) Bud-eating by Bullfinches in relation to the natural food supply. *Journal of Applied Ecology* 1, 265–279.

OWEN, R. W.; PEMBERTON, R. T. (1962) Helminth infection of the starling (*Sturnus vulgaris* L.) in Northern Ireland. *Proceedings of the Zoological Society of London* 139, 557–587.

PALMER, T. K. (1973) The House Finch and Starling in relation to California's agriculture. In Kendeigh, S. C.; Pinowski, J. (*Eds*) Productivity, population dynamics and systematics of granivorous birds. Warsaw; Polish Scientific Publishers. p. 275–290.

PILCHARD, E. I. (1965) Experimental transmission of transmissible gastroenteritis virus by Starlings. *American Journal of Veterinary Research* 26, 1177–1179.

PLANT, C. W. (1978) Salmonellosis in wild birds feeding at sewage treatment works. *Journal of Hygiene,* Cambridge 81, 43–48.

POTTS, G. R. (1967) Urban Starling roosts in the British Isles. *Bird Study* 14, 35–42.

RICKABY, C. D. (1978) A review of the nutritional aspects of complete diets for dairy cows. *Agriculture Development and Advisory Service Quarterly Reviews* 29, 51–70.

ROYALL, W. C. (1975) Starlings feed on ripening sorghum at Cheyenne Bottoms, Kansas. *Kansas Ornithological Society Bulletin* 26, 10.

SEMYA, A. (1976) Starling damage in vineyards of south-east Kazakhstan and analysis of acoustic repulsion results. *Ornitologiya* 12, 160–165.

SHARROCK, J. T. R. (1976) The Atlas of Breeding Birds in Britain and Ireland. Tring; British Trust for Ornithology.

SHI-CHUN, L.; XI-YUO, L.; YAO-KUANG, T.; YAU-KUA, S. (1975) On feeding habits of the Rose-coloured Starling (*Sturnus roseus*) and its effects on lucust-population density. *Acta Zoologica Sinica.* 21, 71–77.

SNOW, D. W. (1968) Birds and the 1967–68 foot-and-mouth epidemic. *Bird Study* 15, 184–190.

STEINBACHER, J. (1960) Die tragödie der Stare von Tunisien. *Gefiederte Welt* 8, 148–150.

STICKLEY, A. R.; DOLBEER, R. A.; WHITE, S. B. (1976) Starling damage to sprouting wheat in Tennessee. *Proceedings of the Bird Control Seminar* 7, 30–38.

STONE, C. P. (1973) Bird damage to tart cherries in Michigan, 1972. *Proceedings of the Bird Control Seminar* 6, 19–23.

SVENSSON, L. (1975) Identification guide to European passerines. Stockholm; Naturhistoriska Riksmuseet.

TAHON, J. (1973) Request for an IBP subgroup "*Sturnus vulgaris* L." to study the biology of Starlings and to control the damage they cause; with a discussion of the effects of different methods of disturbance at the roost on behaviour of Starlings. In Kendeigh, S. C.; Pinowski, J. (*Eds*) Productivity, population dynamics and systematics of granivorous birds. Warsaw; Polish Scientific Publishers. p. 291–297.

TAHON, J. (1978) L'Etourneau Sansonnet (*Sturnus vulgaris*). Biologie, dortoirs, protection des cultures. *European Plant Protection Organisation Publications Series B* 84, 83–153.

TOSH, F. E.; DOTO, I. L.; BEECHER, S. B.; CHIN, T. D. Y. (1970) Relationship of Starling-Blackbird roosts and endemic histoplasmosis. *American Review of Respiratory Disease* 101, 283–288.

VAN DER MUEREN, E. (1977) Dominance hierarchy in a group of caged Starlings *Sturnus vulgaris*. *Le Gerfaut* 67, 267–270.

WARD, P. (1965) Feeding ecology of the Black-faced Dioch *Quelea quelea* in Nigeria. *Ibis*. 107, 173–214.

WARD, P. (1973) A new strategy for the control of damage by queleas. *Pest Articles and News Summaries* 19, 97–106.

WARD, P. (1977) Fat and protein reserves of Starlings. *Institute of Terrestrial Ecology Annual Report*. p. 54–56.

WEINS, J. A.; JOHNSTON, R. F. (1977) Adaptive correlates of granivory in birds. In Pinowski, J.; Kendeigh, S. C. (*Eds*) Granivorous birds in ecosystems. Cambridge; University Press. p. 301–340.

WEST, R. R. (1968) Reduction of a winter Starling population by baiting its pre-roosting areas. *Journal of Wildlife Management* 32, 637–640.

WHISTLER, H. (1949) Popular handbook of Indian Birds. Edinburgh; Oliver and Boyd.

WILSON, J. E.; MACDONALD, J. W. (1967) Salmonella infection in wild birds. *British Veterinary Journal* 123, 212–219.

WITHERBY, H. F.; JOURDAIN, F. C. R.; TICEHURST, N.; TUCKER, B. W. (1940) The Handbook of British Birds. Volume 1. London; Witherby.

WRIGHT, E. N. (1973) Experiments to control Starling damage at intensive animal husbandry units. *European Plant Protection Organisation Bulletin* 9, 85–89.

ZELENY, L. (1969) Starlings versus native cavity-nesting birds. *Atlantic Naturalist* 24, 158–161.

Attempts to control starlings at roosts using explosives

J. Tahon
Ministère de l'Agriculture
Station de Zoologie Appliquée
Belgium

Summary

*From 1972 to 1978, 22 starling (*Sturnus vulgaris*) roosts have been destroyed using explosives during July in order to protect the Sint-Truiden cherry crop. In a roost, almost all the starlings were killed using dynamite. Over seven years, nearly 750 000 starlings have been destroyed. Around 500 000 starlings live in the area in July. Juveniles are much more numerous than adults in some roosts but not in others. Analysis of mean recapture indexes of starlings ringed at the nest indicates a large immigration from adjacent areas (100 km and more). Starlings born in the Sint-Truiden area also disperse widely.*

Short-term efficacy of explosions experienced in Belgium is limited to the second half of the picking period and is restricted to part of the starling population. Roost destruction has little medium or long-term influence on starling populations living in the area at the time of cherry ripening. Data are required to assess the relation between the economic efficiency of explosions in reducing damage and costs of these control measures.

Introduction

Cherry-growing is an important fruit industry in the region of Sint-Truiden (Saint-Trond), the main area of commercial cherry-growing in Belgium (Baudewijn 1975).

The damage caused by starlings (*Sturnus vulgaris*) to sweet cherries is one of the major factors endangering this crop. As a result of the combined action of inclement weather, fungus diseases and birds, some growers sometimes pick only half of the potential crop of susceptible varieties. Some cherry trees are completely stripped of their cherries by starlings, especially when an orchard is located near a starling roost. In those circumstances, significant losses occur during the two hours preceding twilight as tens of thousands of birds progressively assemble before going to roost.

During the last 10 years, we have, in July, periodically censused the starlings roosting within 50 km around Sint-Truiden. About half a million starlings have been estimated each year, the birds being distributed in 10 to 20 roosts. We do not yet know what proportion of the local population feeds in orchards, nor how many cherries an individual eats or injures per day. It is not, therefore, possible to assess the damage done by starlings from bird feeding data. We are also unable to estimate the damage from data on crop yields since these involve considerable local variation in expectation and losses, due both to birds and other factors.

For a long while, more or less sophisticated devices, based on pyrotechnics or phono-acoustics, have been used in attempts to deter starlings from feeding in orchards.

Several of these techniques give satisfactory results for at least a few days. In order to be effective, however, most of them require the presence of operators from dawn to dusk, leading to high costs. As a result, growers make less and less use of them, in spite of their relative efficacy.

The Services of the Ministry of Agriculture have therefore been asked by the "Veilingen" ("Growers Professional Corporation" or "Dutch Auction Ltd") to organise official starling control, with the main efforts to be focused upon the roosts, where all the birds congregate. Hunters were first invited to shoot starlings at the roosts, but this was not effective. To remove birds from roosts, pyro-acoustic devices and "AV Alarm" were then successfully and easily used. However, this failed to prevent damage in orchards, and growers refused their help and did not report the whereabouts of replacement roosts that they discovered in the 3 000 sq km zone. As a substitute to these techniques, the growers then asked the Ministry to dynamite the roosts.

The technique of dynamiting a roost

As soon as cherries of the earliest varieties ripen, in early June, flocks of 10 to 100 birds gather and stay in orchards and their surroundings. The flocks progressively increase in size and number. At first, birds sleep in numerous roosts in small thickets of some hectares scattered around the country but soon birds from the small roosts progressively gather together to form fewer but larger ones. Inquiries about the positions of roosts begin in mid-June. At dusk, between 2000 h and 2200 h, 15 specialists of the Plant Protection Service of the Ministry of Agriculture try to follow flight lines of starling flocks and to localise the roosts. At the same time growers and local inhabitants are invited to transmit similar data to the "Veilingen" where all the records are plotted.

During the period when roosts are small and scattered, they are unstable and no action could be undertaken to destroy them, but by the end of June, some have considerably increased in size and starlings are more and more faithful to them. From then on, it is feasible to dynamite these roosts. As soon as the several licences needed for an operation are received, firemakers, with help of the Plant Protection Service Staff, fix the framework of the explosion. Strict surveillance of the roost is then maintained by policemen, foresters and gamekeepers.

During the evenings prior to the explosion, the specialists of the Plant Protection Service explore all bushes and woods existing as far as 10 km around the thicket to be dynamited. If they detect starlings, they frighten them by voice and by firing shots using pistols in an attempt to concentrate the birds into the target thicket. At the same time, three or four experienced assistants, discreetly scattered a few hundred metres from the thicket, try to estimate as accurately as possible the size of every starling flock joining the roost. With practise, they are able to keep estimation errors under 15% of the total number.

Organisers first mark out in the thicket the place to be dynamited. It is an area of some tenths of a hectare (e.g., 40×60 m), frequently located in a corner of the thicket and densely crowded by at least 20 to 30 000 starlings. During summer, as a substitute for their original reed-bed roosts, starlings always select dense thickets, sometimes under timber. A strip of ground one metre wide is cleared at the circumference and a chicken-wire fence 100 cm high is erected around the thicket in order to avoid the escape of starlings injured by the explosion. A narrow path is then cut into the middle of the roost and the main line of firing cord is unrolled. Secondary lines connected with the main line at 3 m intervals are drawn from the centre to the wire fence. Finally, lines fastened to the highest branches run down every 3 m to the secondary lines. These vertical lines hold 100 g (or 250 g) of dynamite charges hung at each 3 m, for example at 8 m, 5m and 2 m high.

As a rule, the firemakers hang up a total of 200 to 300 kg of dynamite in a network where charges are present every 3 m in each direction.

Charges are always attached to valueless shrubs. A strip of shrubs 3 m wide at the outside edge of the thicket is not dynamited in order to avoid damage to buildings within a radius of 2 km.

The main line is extended some hundred metres away from the thicket and connected at the last moment to a detonator.

The whole task is carried out in a three-day period by 40 firemakers and by the 15 specialists of the Plant Protection Service. As soon as the dynamite is in place, a guard prevents any human access to the thicket.

Detonation is generally planned at 2300 h but is sometimes delayed till 0200 h when too many sightseers are present. In a radius of 2 to 3 km, inhabitants are advised to keep their windows open.

A quarter of an hour before detonation, after dark and when the roost is becoming quieter, 10 people enter slowly into the thicket at the remotest point from the roost. Knocking trees with a stick, shining lights on the tree tops and speaking loud, they concentrate all the starlings in the area with explosive charges. As long as the noise level is not too high, starlings do not leave the thicket. As soon as the starlings are concentrated in the explosion area, these people leave and the charges are detonated. Immediately after the explosion firemakers check all the charges and a Civil Saveguard Unit illuminates the wood with floodlights.

Some starlings are not killed outright by the explosion and 20 men enter the dynamited area and kill these birds within two or three hours. Then over one or two days, two teams of 15 collectors collect the dead birds and put them into bags which are brought into "Veilingen"refrigerator chambers. Seven to 10 scientists then sort about 15 000 birds a day into two categories (adults and juveniles), count them and collect every ring (one or two rings per hundred birds). Carcasses are incinerated.

As a rule, most of the starlings sleeping in a roost are killed, only a few being uninjured and able to fly away. The roosts do not generally house other bird species except some possible tens of house sparrow (*Passer domesticus*) and of tree sparrow (*Passer montanus*). Dynamited roosts are not reoccupied for two years or more.

Roost destruction experiments sometimes are unsuccessful: this is due to an unexpected absence of most of the starlings, or to a failure to concentrate the birds into the right area. But in any case, for security rules, charges must be exploded as planned.

The number of birds killed

The geographical locations of the 22 explosions carried out from 1972 to 1978 are shown in Fig. 1 and the numbers of birds killed are given in Table 1. The number of roosts dynamited each year is not dependent on the number of known roosts nor on the number of birds living in the area, but is restricted by technical considerations.

During the seven seasons, 749 000 starlings have been killed in the 22 explosions carried out, a mean of 107 000 starlings per year, with a maximum of 280 000 in 1975 and a minimum of 28 000 in 1972. As said before, more than half a million starlings are present in the area each summer. The mean number of starlings killed per explosion is 34 000, with a maximum of 100 000 and a minimum of 4 000.

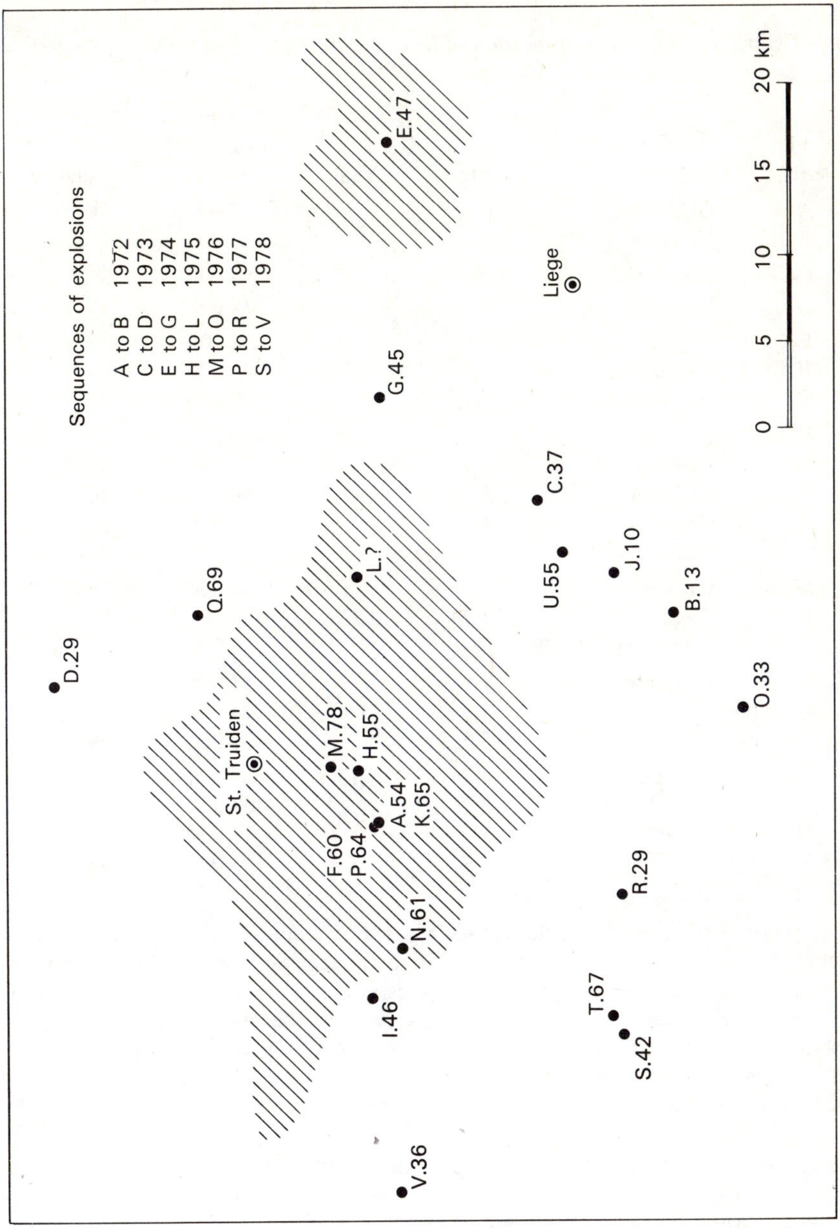

Figure 1. Dynamited roosts and proportion of juveniles in the birds killed.

Table 1
Number of starlings killed and proportion of juveniles in roosts dynamited between 1972 and 1978

Code letter	Locality	Date	Year	Number of killed starlings	% of juveniles
A	Gingelom	19.7	1972	7 000	54
B	Verlaine	25.7		21 000	13
C	Freloux	10.7	1973	15 000	37
D	Stevoort	12.7		16 000	29
E	Visé	25.6		16 000	47
F	Gingelom	2.7	1974	45 000	60
G	Diets-Heur	9.7		70 000	45
H	Muizen	3.7		100 000	55
I	Neerheylissen	5.7		35 000	46
J	Verlaine	8·7	1975	60 000	10
K	Gingelom	11.7		25 000	65
L	Horpmaal	30.7		60 000	?
M	Kerkom	29.6		16 000	78
N	Racour	3.7	1976	4 000	61
O	Vinalmont	6.7		27 000	33
P	Gingelom	7.7		51 000	64
Q	Ulbeek	13.7	1977	24 000	69
R	Avin	15.7		30 000	29
S	Boneffe	4.7		53 000	42
T	Branchon	18.7	1978	31 000	67
U	Jeneffe	25.7		12 000	55
V	Melin	27.7		31 000	36

The proportion of juveniles in the roosts has varied from 10 to 78% (Fig. 1). This proportion depends on various factors: roosts already occupied before reproduction still attract adults after breeding, but juveniles soon establish new roosts. The percentage of adults generally increases as the season goes on, probably because earlier in the season most of them are rearing a second brood and do not join a roost. The percentage of juveniles generally increases in previously dynamited areas. This last observation suggests that there is a large juvenile immigration into the area during the fruit ripening period.

Efficiency of roost destruction in population control: an assessment based on ring recoveries

From 1976 to 1978 the staff of the "Station for Applied Zoology" in Gembloux have erected starling nestboxes in different woods: 1 250 nestboxes in groups of 50 in 1976, 1 280 nestboxes in groups of 80 in 1977 and 1 350 nestboxes in groups of 50 in 1978 (Fig. 2). Nesting data from 1976 and 1977 have already been published elsewhere (Tahon

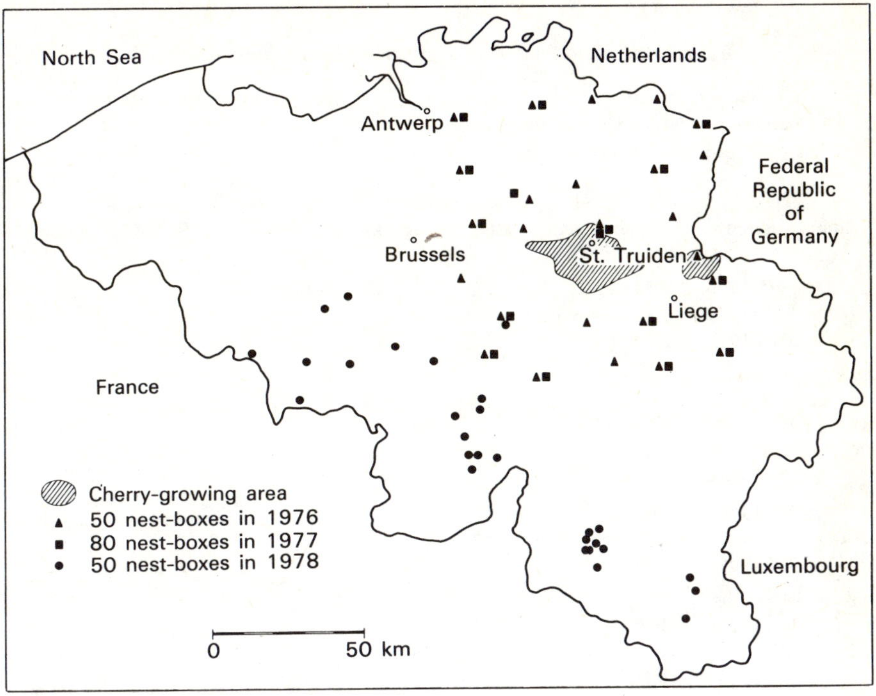

Figure 2. Ringing efforts in 1976–1978.

et al. 1978). Altogether 11 004 pulli were ringed (Table 2); 314 of them died before fledging.

Amongst the 10 690 fledged birds, 6 555 (61·3%) were born from 1 866 clutches laid during the first nesting period (early clutch), 2 346 (21·9%) from 741 clutches laid during the replacement period (intermediate clutch) and 1 789 (16·8%) from 985 clutches laid during the second nesting period (late clutch). One-hundred of these ringed pulli were collected as juveniles amongst the 140 232 young starlings found dead on the ground after the 10 explosions in 1976 to 1978.

Of the ringed birds found dead in the roosts, the proportions from the three periods of reproduction were, respectively, 70, 24 and 6% of the total. In comparison with the proportions from each period that were ringed, second brood juveniles appear to escape proportionately better from explosions than the other broods. This is because some of these pulli are still in the nest at the time of the explosions.

Knowledge of the exact birth and ringing places of all these pulli, as well as of the exact location of the explosion (Fig. 3), makes it possible to establish recapture indexes and to group them into classes of distance. A recapture index (I) is a function of the recapture percentage $\frac{n_{d(r)}}{n_r}$ (the number of dynamited birds ringed ($n_{d(r)}$) divided by the number of pulli ringed and not previously reported dead (n_r) and is inversely related to

Table 2

Nest ringing data and ring recoveries in dynamited roosts

1976 → 25 localities, with 50 nestboxes = 1 250 nestboxes
1977 → 16 localities, with 80 nestboxes = 1 280 nestboxes
1978 → 27 localities, with 50 nestboxes = 1 350 nestboxes

Ringed juveniles recovered at roost

Ringing periods	Pulli ringed		Dead at nest after ringing		1976 29/6 M	1976 3/7 N	1976 6/7 O	1977 7/7 P	1977 13/7 Q	1977 15/7 R	1978 4/7 S	1978 18/7 T	1978 25/7 U	1978 27/7 V	Total	
1st brood																
10/5 to 19/5 1976	2 591		72		17	2	14								33	
02/5 to 19/5 1977	1 824	6 685	24	130				15	10	2					27	70
18/5 to 31/5 1978	2 270		34								4	3	1	2	10	
Replacement brood																
19/5 to 15/6 1976	500		14		9	1	0								10	
19/5 to 20/6 1977	1 295	2 410	32	64				5	6	2					13	24
20/5 to 20/6 1978	615		18								1	0	0	0	1	
2nd brood																
15/6 to 9/7 1976	706		37		0	0	0								0	
20/6 to 9/7 1977	772	1 909	47	120				1	5	0					6	6
22/6 to 9/7 1978	431		36								0	0	0	0	0	
Total																
1976	3 797		123		26	3	14								43	
02/5 to 9/7 1977	3 891	11 004	103	314				21	21	4					46	100
1978	3 316		88								5	3	1	2	11	
Number of juveniles killed in explosions					11 897	2 192	8 806	31 857	16 364	8 369	22 189	20 522	6 850	11 186		
					22 895			56 590			60 747					
								140 232								

the number of dynamited birds (n_d). To produce an easy statement the result is multiplied by 10^7. The formula is written:

$$I = \frac{n_{d(r)} 10^7}{n_r \, n_d}$$

A different recapture index is calculated for juveniles and adults.

In practice, (see the example in Fig. 4 and Table 3) for each explosion all the vectors from the roost to the different ringing points are drawn in order to measure the exact distance between ringing and recatching (explosion) places. Each vector is then set in a 10 km distance class. Summing all the vectors from all the explosions, the number of vectors to be studied in each distance class is obtained. In each class, the mean recapture index ($\bar{I}$) is then calculated as the sum of all the recapture indexes of that class (ΣI), divided by the number of vectors of the class (n_I). The mean recapture index (I) is: $\frac{\Sigma I}{n_I}$

Data collected in 1976–78 are presented in Table 4. As expected, mean recapture indexes decrease when distance between the ringing point and the explosion point increase. For example, there is a ten times higher probability of recovering a bird ringed 15 km from a dynamited roost than of recovering a starling ringed 45 km away. The ratio between the

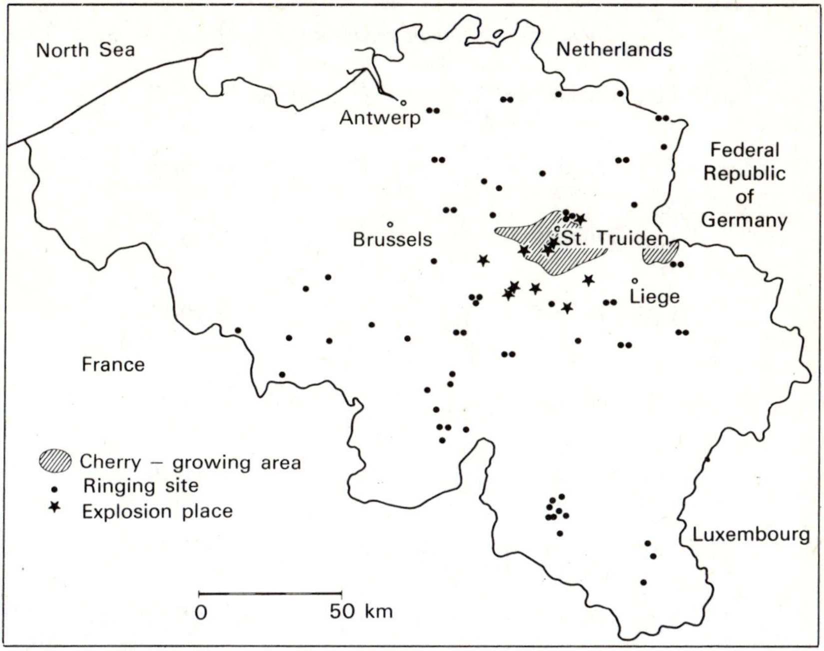

Figure 3. Ringing sites and explosion places – period 1976–1978.

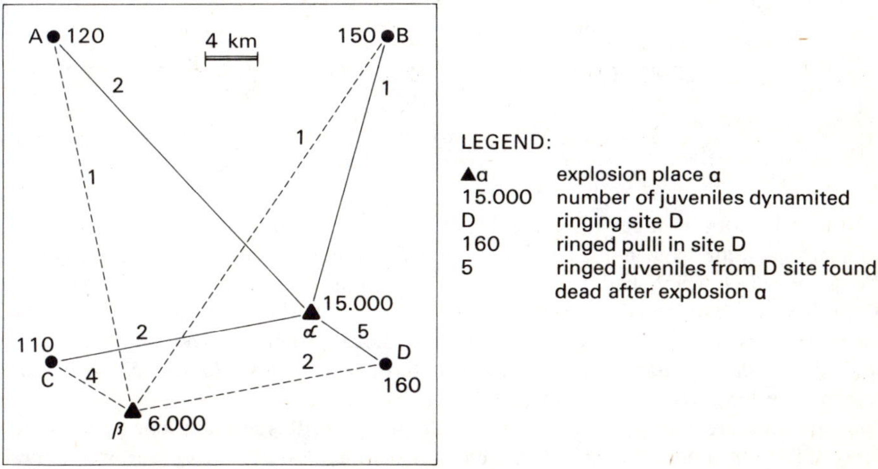

Figure 4. Calculation of mean recapture indexes ($\bar{I}$).

Table 3

Calculation of mean recapture indexes ($\bar{I}$)

$$I_{\alpha A}\ (30\ \text{km}) = \frac{2}{120} \times \frac{10^7}{15\,000} \text{ or } K_\alpha = 11{\cdot}11$$

$$I_{\alpha B}\ (22\ \text{km}) = \frac{1}{150} \times K_\alpha = 4{\cdot}44$$

$$I_{\alpha C}\ (20\ \text{km}) = \frac{2}{110} \times K_\alpha = 12{\cdot}12$$

$$I_{\alpha D}\ (8\ \text{km}) = \frac{5}{160} \times K_\alpha = 20{\cdot}83$$

$$I_{\beta A}\ (30\ \text{km}) = \frac{1}{118} \times \frac{10^7}{6\,000} \text{ or } K_\beta = 14{\cdot}13$$

$$I_{\beta B}\ (36\ \text{km}) = \frac{1}{149} \times K_\beta = 11{\cdot}19$$

$$I_{\beta C}\ (8\ \text{km}) = \frac{4}{108} \times K_\beta = 61{\cdot}73$$

$$I_{\beta D}\ (20\ \text{km}) = \frac{2}{155} \times K_\beta = 21{\cdot}51$$

km	10	20	30	40
n b of vectors	2	3	2	1
$\frac{\text{n b dynamited}}{\text{n b ringed}}$	$\frac{9}{268}$	$\frac{5}{415}$	$\frac{3}{238}$	$\frac{1}{149}$
%	3·36	1·20	1·26	0·67
I	$\frac{20{\cdot}83+61{\cdot}73}{2}$	$\frac{4{\cdot}44+12{\cdot}12+21{\cdot}51}{3}$	$\frac{11{\cdot}11+14{\cdot}13}{2}$	$\frac{11{\cdot}19}{1}$
I	41·28	12·69	12·62	11·19

areas of circles with these radii is 1 to 3. Due to these two opposite factors, the numbers of starlings originating from more and more remote distance classes decrease only slowly.

Some juveniles which had been ringed two months earlier, at distances exceeding 100 km from the dynamited roost, have been recovered. On the other hand, relatively few birds ringed close to the roost have been recorded. Local juveniles would not be especially attracted to cherry orchards surrounding their birthplace, but would also disperse some distance. The long-term efficacy of explosions decreases as well.

Similar recapture indexes are shown for adult birds ringed as nestlings one or two years before, indicating similar movements for birds of both age classes. Moreover, if a natural mortality rate of 66% is accepted for the first year birds (Delvingt 1962), only one-third of

Table 4

Mean recapture indexes

A – juveniles

	A	B	C	D	E	F	G	H	I	J	K	L	M	N
Distance in km	0–10	11–20	21–30	31–40	41–50	51–60	61–70	71–80	81–90	91–100	101–110	111–120	121–130	131–140
Number of vectors	3	8	15	31	46	38	26	29	22	12	11	3	3	1
ringed/recovered	15/191	23/449	11/1 431	11/3 663	18/6 537	15/6 684	4/4 495	1/4 197	1/2 541	1/1 039	0/591	0/363	0/182	0/100
Percentage	7·85	5·12	0·77	0·30	0·28	0·22	0·09	0·02	0·04	0·09	0	0	0	0
Mean recapture indexes	617·2	307·6	161·9	58·3	31·6	22·6	13·2	2·0	2·9	7·4	0	0	0	0

B – adults

	A	B	C	D	E	F	G	H	I
Distance in km	0–10	11–20	21–30	31–40	41–50	51–60	61–70	71–80	81–90
Number of vectors	1	14	41	50	41	44	31	22	13
ringed/recovered	1/19	34/1 656	41/5 361	20/6 274	9/6 544	6/8 042	3/5 410	1/4 863	2/3 126
Percentage	5·26	2·05	0·76	0·32	0·14	0·07	0·06	0·02	0·06
Mean recapture indexes	698·2	180·7	68·0	24·5	13·0	6·4	5·1	1·5	5·5

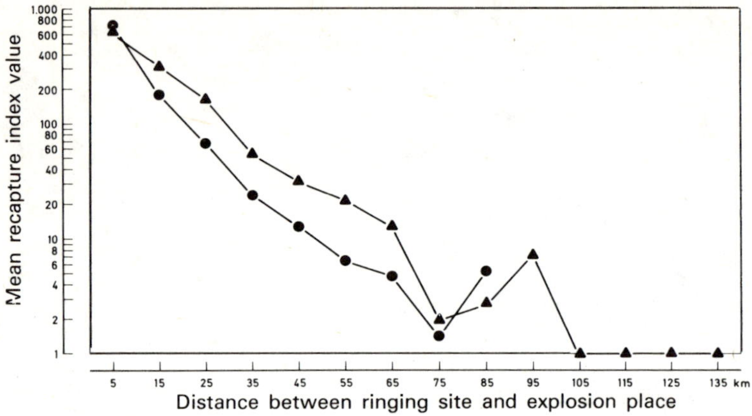

Figure 5. Comparison between mean recapture indexes of juveniles and adults.

the birds are still alive after one year, and it is quite normal to observe, as in Fig. 5, lower mean recapture indexes for adults. It is also obvious from the parallel data of the two curves in the three or four shortest distance classes that the local population has not suffered badly from the mortality due to explosions. If a large number of local juveniles had been killed in explosions, the curve for adult birds (Fig. 5) should have started near zero, increased to 25–35 km, and then decreased further away. This partly explains the inefficiency of the killing of starlings in roosts as a form of population control.

Discussion and conclusions

The sweet cherry-growing carried out in the Sint-Truiden region is an important industry for many local fruit-growers and it employs large numbers of pickers. However, many unfavourable factors (rise in cost of land and labour, unsteady income due to climatic factors, virus diseases, lack of productivity during the first 15 years of tree life, damage caused by fungus diseases and birds) have contributed to a decline in the industry.

Predation by starlings is one of the few factors against which control is possible. Therefore, for psychological reasons, efforts to control starlings have been officially and *a priori* managed and maintained. Dynamiting roosts is the most spectacular technique used so far and its efficiency can now be discussed in the light of seven years' experience.

Short-term results

The starling population (500 000 birds during the cherry season in that region) is reduced by the number of killed birds (100 000 starlings). This gives some short-term protection covering the second part of the cherry crop. Unfortunately, roosts only grow in size at the end of June, precluding earlier control and leaving early cherry varieties unprotected.

As a consequence of the explosions, starling roosts move progressively away from the cherry-growing area. This leads to a reduction in the damage previously done by pre-roost assemblies within a 2 to 3 km radius around the roosts. This decrease in the most spectacular type of damage is readily apparent. New summer roosts are, however, formed in less accessible places, sometimes even in towns or in nature reserves.

Long-term results
According to advocates of roost destruction, starlings causing damage to cherries belong to local populations, having bred or been born less than 30 km away from the orchards where they feed. According to others, some of the starlings come from further away, in unknown proportions.

We have shown that 500 000 starlings are present in the area in July. The local breeding population (30 km around Sint-Truiden) is evaluated to 25 000 pairs, on the basis of provisional data collected for the "Atlas of breeding birds of Belgium". About six fledglings are raised by a pair each year as observed in our trials carried out from 1976 to 1978 (see above). After the breeding season, therefore, the local population would not account for more than 50 000 adults and 150 000 juveniles, a quarter or a third of which would still be nesting or present as pulli in the nest during the control period.

The promoters of the use of explosives predicted a severe reduction in the starling population after one year, so that it should have been possible after some time to limit or to stop roost destruction. But from the start, it was evident to biologists that a 5 to 20% reduction of the population was insufficient to gain such a result. Evidence of that misjudgement is provided by the relatively stable number of birds killed each year in a population which is still approximately at the same level as before.

Movements of starlings and control
Knowledge of starling movements is of prime importance when planning control strategies. If starlings move only a little (some km) but their populations are not severely reduced by destruction of their roosts, either because some roosts remain unknown, or because some known roosts are not dynamited, control of the local population is insufficient to reduce damage. If starlings wander more widely (some tens of km), the area where the roosts are to be dynamited must be enlarged, with a proportional increase in number of explosions and cost. Finally, if the area where starlings have been eliminated by roost destruction is rapidly reinvaded by immigrant birds, such a control offers no long-term solution. Our data show that all three kinds of movement are involved.

Ringing results demonstrate that juveniles rapidly move rather far away from nest. Some birds leave the cherry-growing area, while others enter it and around 20% of the juveniles killed in the roosts were birds born in each of the distance classes: 0 to 10 km, 10 to 30 km, 30 to 60 km, 60 to 100 km and more than 100 km. Similar results have been obtained for adult birds, although there are far fewer of these roosting in the cherry-growing area June–July.

Roost destruction in the cherry-growing area has little influence on other Belgian starling populations, nor does it really modify the local population level regularly attained in other seasons: several million starlings migrate from September to December through the Sint-Truiden zone, some of them staying there throughout the winter.

Conclusions
Techniques for killing starlings using explosives have been investigated in detail in Belgium, but their efficiency is limited to reducing damage in the second half of the picking period and restricted to only a part of the starling population.

Staff costs (more than 100 persons actively participate) are impossible to estimate because several organisations are involved in these actions. Economic losses caused by starlings to cherry crops have not yet been accurately estimated in Belgium and therefore the relationship between economic efficiency of explosives as a means of control and costs of these actions remain practically unknown.

Acknowledgements

This investigation was conducted under the direction of J. Bernard, Directeur de la Station de Zoologie appliquée de l'Etat à Gembloux, who regularly in the course of the research helped with valuable and stimulating concepts and provided large work facilities. Many thanks are due for that beneficial support.

The author is also very much indebted to C. J. Feare for revising the English text.

References

BAUDEWIJN, J. E. A. (1975) Belang van de Kersenteelt. *Coördinatie Comite Bescherming Vogels (CCBV – CCPO),* 3, 6–7.

DELVINGT, W. (1962) L'étourneau en Belgique: longévité, pontes. *Le Gerfaut,* 52, 586–601.

TAHON, J.; TORREKENS, CH.; GIGOT, J. (1978) Données récoltées lors du baguage au nichoir de cultures. *Publications Organisation Européenne et méditerranéenne pour la Protection des Plantes (O.E.P.P.), Série B,* 84, 83–153.

TAHON, J.; TORREKENS, CH.; GIGOT, J. (1978) Données récoltées lors du baguage au nichoir de pulli d'étourneaux sansonnets (*Sturnus vulgaris*) en 1976 et 1977 en Belgique. *Parasitica,* 34, 122–182.

Breeding strategies of the starling

Rudolf Frans Verheyen
Departement Biologie
Universitaire Instelling Antwerpen
Belgium

Summary

A study of the reproductive biology of starlings at Brecht in Belgium, and a synthesis of the most relevant literature, provide evidence that Sturnus vulgaris *possesses several behavioural mechanisms that maximise reproductive output. The selective advantage of exploratory behaviour during autumn and spring, and the choice, possession and defence of suitable nest holes against conspecifics are discussed. Lack of nest site tenacity, lack of mate fidelity, and the regular occurrence of simultaneous and successive polygyny seem to be advantageous for the reproductive success. Changes in the behaviour of male and female members of a pair, in the absence of the partner, are described. Colonial breeding seems to be advantageous since it probably promotes early egg-laying, through social stimulation, and polygyny. No significant correlation between colony size and synchrony has been found; however, larger colonies have a tendency to start egg-laying earlier than smaller ones. The principal factors determining the timing of egg-laying and the number of second clutches are briefly discussed.*

Introduction

Since the middle of the last century the population of starlings (*Sturnus vulgaris*) has increased in many places in Europe and has extended its range. Starlings have also increased explosively where they have been introduced in North America, Australia, New Zealand and South Africa. Gallacher (1978) gives a survey of the expansion of starlings. Two critical questions arise: (1) What are the population regulating mechanisms that have permitted an increase in numbers and range and (2) Why are starlings so successful in comparison with other hole nesting birds?

According to Perrins (1970), natural selection will promote early breeding so that young are in the nest at the best time for raising them, while external factors, e.g., food supply, will counteract. Crossner (1977) suggested that starling populations are reproducing maximally: the mean clutch-size in an area is close to, but less than, the optimal clutch size, so that the probability of survival of the young in the nest and after fledging is high, and also the body condition of the parents remains optimal, leading to a better chance of raising a subsequent brood.

In this paper I shall discuss some behavioural mechanisms that may affect reproductive output.

Nest site occupation and pair formation

In theory, it is important for starlings to look for nest cavities early, and to occupy and keep the most suitable ones, so that pairs can be formed with as many successful partners as possible.

The selection of a suitable nest hole is, of course, very important for the survival of the brood: a bad choice may cause the complete or partial loss of the brood, e.g., by predation (crow, squirrel), by men (trapping of adults on the nest, destruction of the nest), or by the environment (e.g., rainfall, humidity in the nest, sunshine). Verheyen (1969) demonstrated that the elevation of the nest hole above the ground, and the orientation of the entrance of the hole (probably together with other factors), play an important role in determining the birds' choice, with high cavities facing east or south-east being preferred.

Starling populations show important regional differences in hole visitation: British and American starlings visit their nest holes throughout the year, with a lower frequency during July and August (Morley 1941, Bullough 1942, Kessel 1957, Troetschler 1976), while continental starlings do so only during the breeding season, and during a short period in September/October (Kluyver 1935, Berthold 1964). Territorial behaviour and pair formation in the British and American populations may occur in September/October and from January, starlings of two-years-old and older, may roost together at night in the nest cavities. The number of birds sleeping in holes increases quickly so that from February/March the important roosts are visited principally by yearling birds (Wynne-Edwards 1929, 1931, Marples 1932, 1934).

In our study area in Belgium (51°N, 4°E), hole visitation, territorial behaviour and pair formation generally begin in February and increase considerably during March. Kluyver's (1933, 1935) research in Wageningen (52°N, 5°E), and my own observations, suggest that the oldest males visit the cavities from the middle of February onwards and occupy the most suitable nest holes. Their territory is relatively large at first and embraces other cavities besides the most suitable "principal nest hole" (Wallraff 1953). Somewhat later (February–March), the two-year-old males, and the two-year and older females, arrive at the colony: the recently arrived males take and occupy one or more cavities belonging to the already settled, older males, those taken being probably less suitable than the "principal nest hole" of the previous owner. At first the recently arrived two-year-old or older females visit many nest holes and select the most suitable, which are mostly already occupied by the two-year-old and older males. In spite of the limited data available, it appears that nest visitation early in spring allows older males to mate with older females and take possession of the most suitable nest cavities, but more data are required to confirm this.

Yearling males and females generally arrive at the colony later: according to Kluyver (1933), 60% of the yearling males and 38% of the yearling females, which are captured in nest boxes, are present in the colony by 15 April; 68% of the yearling females which visit the colony, but only a very small number of yearling males, actually breed (Verheyen 1969). Yearling males almost never reproduce while yearling females do, even though the females arrive at the nest holes later.

Field observations and automatic nest recordings show that most starlings visit the colony immediately after leaving the roost, often even before sunrise, and before feeding, and that the owner(s) of holes are almost always present when intruders visit the colony. It is also shown that there are distinct periods of reproductive behaviour (hole-visitation, fighting, pair formation, nest-building) and of maintenance activities (preening, feeding, rest) (Verheyen 1968–69). During the period of nest site acquisition, starlings utilise specific behaviour mechanisms to chase away competing intruders and to incite potential partners to stay. The activities and postures involved have already been described by many authors (e.g., Morley 1941, Schuz 1942, Wallraff 1953, Schneider 1960, Delvingt 1960–61, Davis 1963, Ellis 1966, Verheyen 1969, Van der Mueren 1977).

The male waves his wings and sings in a typical way ("screaming" according to Davis 1959) from a visible position next to the nest hole. When a starling arrives and settles close to the male, he stops singing immediately: if the intruder is a female, the unpaired male shows his nest hole by flying into it, followed by intensively singing in his hole (Schuz 1943, Wallraff 1953, Verheyen 1969). This "indicative reaction" of the male, followed by nest-visiting of the female, is the start of pair formation in which the nest hole itself plays a prominent role. After mating, the main activity of the pair consists of nest-building and defence of the territory against intruders: male and female recognise each other at a long distance (Delvingt 1960–61, Verheyen 1968–69). Pairs already established at the end of February or beginning of March may remain together; mostly, however, mating occurs later in spring, sometimes only a few days before egg-laying (Verheyen 1968–69). Before mating, the male dominates all intruders, but after mating and until fledging of the first brood, the female dominates the male within the nest hole. Outside the nest hole, the female dominates the male only during the period of nest-building and egg-laying, when the male shows a tendency to flee and "branch-running" may occur (Morley 1941).

Behaviour during the absence of the partner

1. *From the start of mating until egg-laying.* We have frequently observed that during the absence of his mate even for a few minutes, a male will react with conspicuous gamosematic display and show his nest hole to any interested female. Prolonged absence of the mate causes rapid pair formation with a new female, probably on the same day. A short absence of the male does not break down the pair bond, but prolonged absence probably causes the desertion of the nest by the female, followed by rapid mating with a competing male.

2. *From egg-laying until fledging of the first brood.* Disturbance of the brood during this period may result in desertion of the nest by one or both parents. We have twice seen that when the male deserts the nest, the female continues the parental care by herself; eggs hatched normally, but mortality among the pulli was higher. When the female deserts the brood, the male usually proceeds with parental care, i.e., incubation of the eggs, or feeding and brooding the young, mostly during the morning until 1000 h and during the afternoon from 1600 h until the departure to the roost. After a period of time, varying from some hours during egg-laying to some days during incubation, the male visits other nest holes with increasing frequency, even outside his territory. exhibits gasmosematic display and starts carrying conspicuous material, e.g., leaves and flowers, into the nest hole to cover the eggs or dead pulli. When no unpaired females are available, or when mating fails, due to the presence of eggs and pulli in the nest cavity, the male may start removing nest material, eggs and pulli from the nest (this was observed twice from a hide). Periods of display are regularly interrupted by periods of nest-cleaning. After a prolonged absence of unpaired females the male may break down his nest completely by devoting more time to nest cleaning and less to display, and finally desert the nest and even leave the territory.

3. *After the fledging of the first brood.* The female accompanies and feeds the fledglings for some hours or days and then starts visiting nest holes again. The male generally cleans his dirty nest intensively by throwing out parts of the nest material, and we have frequently observed that the nest may be cleaned within a few hours after the fledging of

his brood. Afterwards, the male may visit several holes in the neighbourhood, which induces conflict with competing males. He starts displaying intensively for long periods, frequently showing his nest when a female is present: his tendency to mate is very high. Since the unpaired females also show a strong tendency to visit nest holes and to mate, there is a high probability that pair formation will occur very soon after the fledging of the brood, even in a different nest hole. Prolonged absence of his first female, through feeding her young or visiting nest holes elsewhere, may result in the male mating with a new female. We observed once that a male followed his previous female into the territory of another male more than 50 m away and, in spite of attacks by the competing male, tried to drive the female back to his nest hole.

4. *Mate fidelity and nest-site tenacity.* Constancy of pairing, that is the same pair breeding together in the same nest hole in successive years, appears to be very rare (Kluyver 1933, Kessel 1957) and separation of the partners, even during the same breeding season, frequently happens (Schneider 1957, 1972, Feare & Burnham 1978).

In our study area, change of nest holes during the same breeding season seems to be common, with repeat clutches as well as with the second layings. In 1966, 73% of breeding females (n = 96) were captured on the nest and almost all females that started intermediate or second clutches were caught. Three out of the four females captured on the intermediate clutches (repeat clutches) deserted their previous clutch and changed nest sites: the previous history of the fourth female is unknown. Only three females out of 11 (27%) remained in the same nest-box after fledging the first brood and starting a second clutch. According to Yom-Tov *et al.* (1974), nest parasitism seems to be a common phenomenon in the starling but we found no evidence of this.

From the beginning of pair formation the male tries to mate with a new female as soon as his own is absent. Prolonged absence of the female results in the loss of the brood and sometimes in removal of the brood by the male. The female, on the other hand, generally continues the care of eggs and young when her male is absent. Obviously, there is a conflict of interests between the sexes. It is more advantageous for the breeding success of the female to maintain a suitable nest hole from the start of egg-laying and to raise the brood by herself, since the male has only a small role in parental care (Orians 1972, Askenmo 1977). On the other hand, it appears to be more advantageous for the reproductive success of the male to start a new brood with another female rather than raise its previous clutch alone. More research is necessary to check this but the fact that simultaneous and successive polygyny regularly occur in starlings (Echardt 1919, Freitag 1936, 1937, Kessel 1950, Schifferli 1957, Delvingt 1960–61, Verheyen 1969, Schneider 1972, Feare & Burnham 1978) lends support to the hypothesis. Van Haartman (1951) considered that polygyny increased the reproductive success of the male, and Askenmo (1977), who studied the reproductive strategy of the pied flycatcher (*Ficedula hypoleuca*) concluded that: "Although polygyny basically should be disadvantageous to females, they select for it, because acceptance of a polygynous male may secure a coveted piece of a limited resource which is vital for her breeding success."

Feare & Burnham (1978) found that males who kept their females had been more successful in their first brood then males that changed mates. However, the given explanation, that following a first brood which produced few chicks, males are responsible for the breakdown of the pairbond and try to produce a second brood with a new female, requires further research. They stated that: "How males assess the success of their early brood is not known, but it could involve the amount of effort they have to devote to feeding the chicks, i.e., the extent of post-coital investment". Another explanation, which should be investigated, could be that the female of a less successful first brood is unable to start a second brood with her male. This could be because her physiological condition prevents a second egg-laying, due to the onset of gonad regression, or to lack of protein reserves in the flight muscle (Jones & Ward 1976), or because she had to feed and accompany her fledged brood for some period of time, so that the chances of mating with her first male were reduced.

Factors determining the timing of egg-laying

Jones & Ward (1976) have convincingly demonstrated that the level of protein reserves in the flight muscle is the proximate factor that determines the time of egg-laying and the clutch-size in *Quelea quelea*. It seems preferable to refer to the control of breeding through body condition, in particular the size of protein reserves in the flight muscle, rather than regulation by the food supply. These findings and conclusions apply also to starlings (Ward 1977), although it seems to me that these factors must operate in combination with supplementary stimuli to determine the start of egg-laying.

During nest-building, the female probably receives necessary stimuli for the final stage during the growth of the follicles. External stimuli, derived from the behaviour of the nest-building male, from other starlings through social stimulation, from the availability of suitable material and the presence of a cavity with a nest (usually started by the male and always completed by the female), influence the building behaviour of the female directly (Verheyen 1968–69). The development of the incubation patch and the further reproductive development up to egg-laying are probably also promoted by these stimuli (Bissonette & Zujko 1936, Burger 1949, 1953 and Lloyd 1965 for starlings, Hinde 1967 and White & Hinde 1968 for canaries).

Our observations on caged starlings seem to confirm this hypothesis: I twice observed that unpaired females kept alone, but not visually or acoustically isolated, in outdoor aviaries, start egg-laying at the same time as paired females in other outdoor aviaries and as starlings breeding in the wild. On the other hand, the sex drive of the male is probably stimulated by the building behaviour of his female and by the behaviour of the other starlings in the colony.

In my study area, (51°N 4°E), building takes 2–18 days ($\bar{x}$=12, n=376) but in Finland (63°N 23°E) only 3–10 days (x=6·5, n=27) is spent on nest building (Korpimäki 1978). Starlings arriving later at the breeding place in my study area, e.g., yearling females, build faster than the earlier ones, so that most nests are completed at about the same time; social stimulation probably plays an important part in this synchronisation of nest-building. Schwab & Lott (1969) and Gwinner (1975) experimentally demonstrated the effect of social stimuli on the endogeneous, pre-programmed, autumn and spring reproductive activity of the starling.

Egg-laying of the first brood is highly synchronous (Fig. 1) but in our study area this is subject to annual variations ($\bar{x}$: 6,7 days, range 4–10 days, n=8), which have also been shown by other investigators (e.g., Schneider 1957, 1963, Haarhaus 1968, Collins & De Vos 1966, Royall 1966, Tenovuo & Lemmetyinen 1970, Bogucki 1972). Our results did not confirm the "Fraser Darling" hypothesis that large colonies should show a greater

Egg-laying

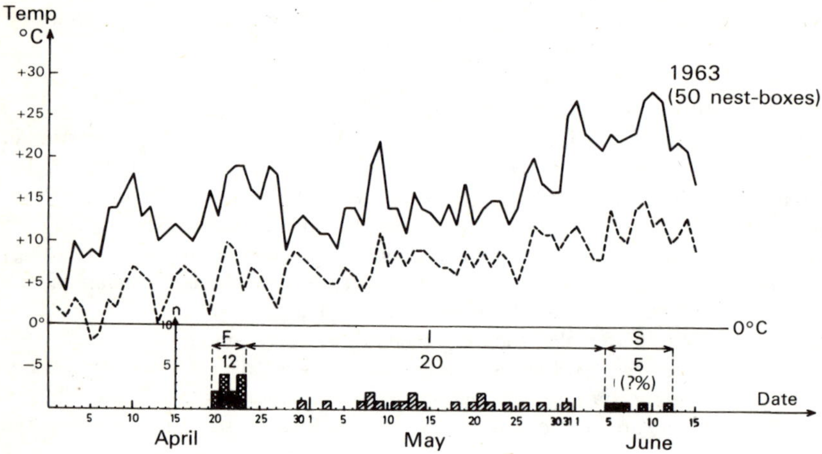
Temp
°C
+30
+25
+20
+15
+10
+5
0°
−5
1963
(50 nest-boxes)
0°C
n
10
5
F
12
I
20
S
5
(?%)
Date
April
May
June

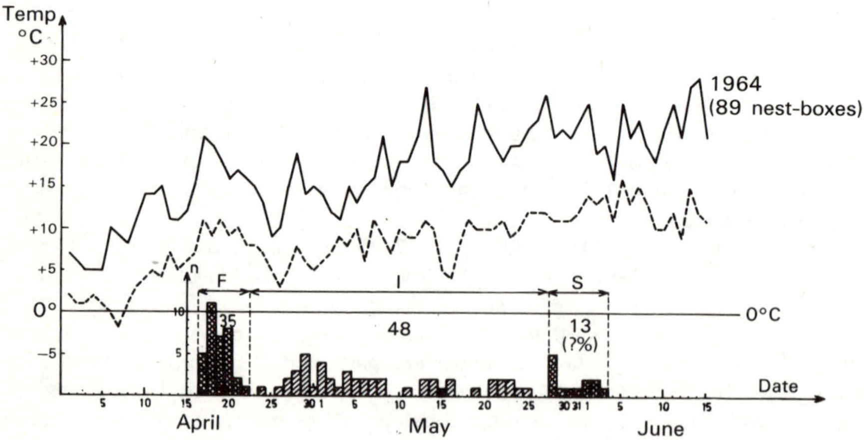
Temp
°C
+30
+25
+20
+15
+10
+5
0°
−5
1964
(89 nest-boxes)
0°C
n
F
35
I
48
S
13
(?%)
Date
April
May
June

Egg-laying

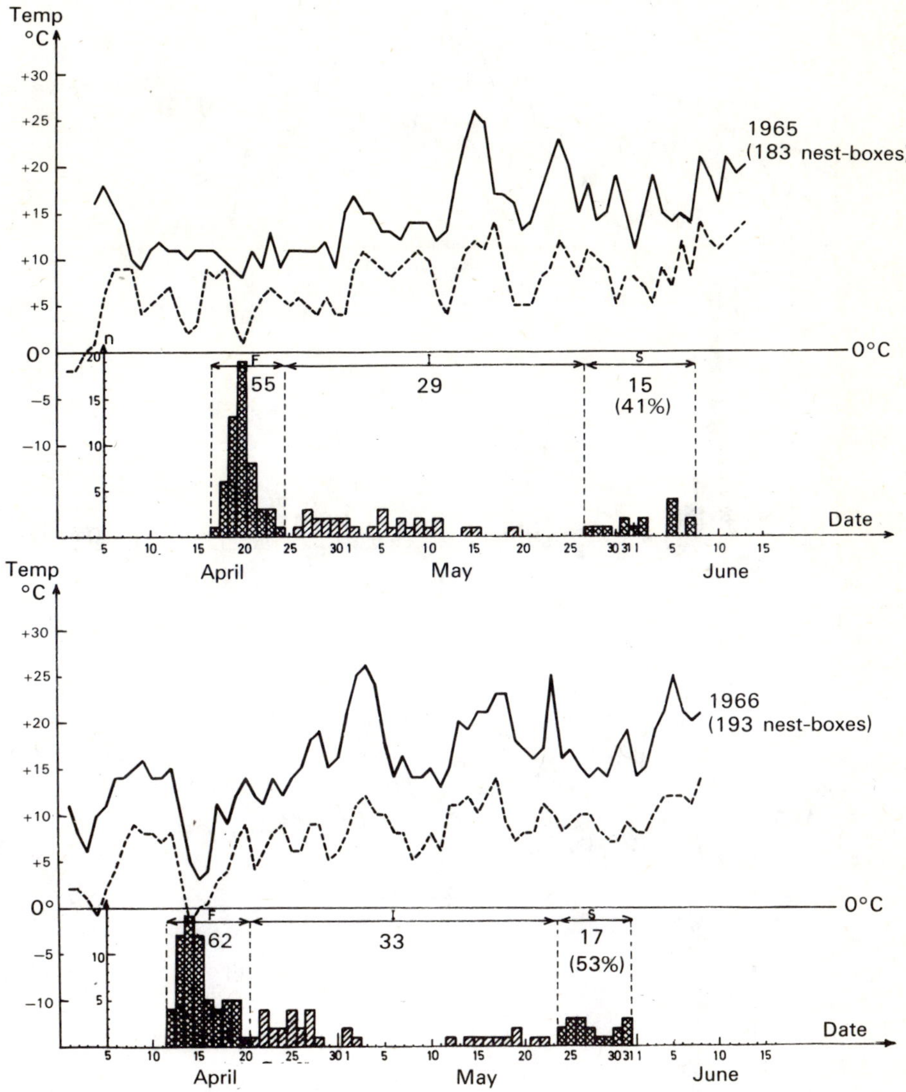

Figure 1. The number of clutches initiated each day (n), and the daily maximum and minimum temperatures, for the breeding seasons 1963 to 1970. F = first clutch, I = intermediate clutch, and S = second clutch. Egg-laying of the first brood is highly synchronous and the start of laying appears to be related to the weather conditions (especially precipitation and temperature between 0500 h and 1000 h) during the week preceding egg-laying.

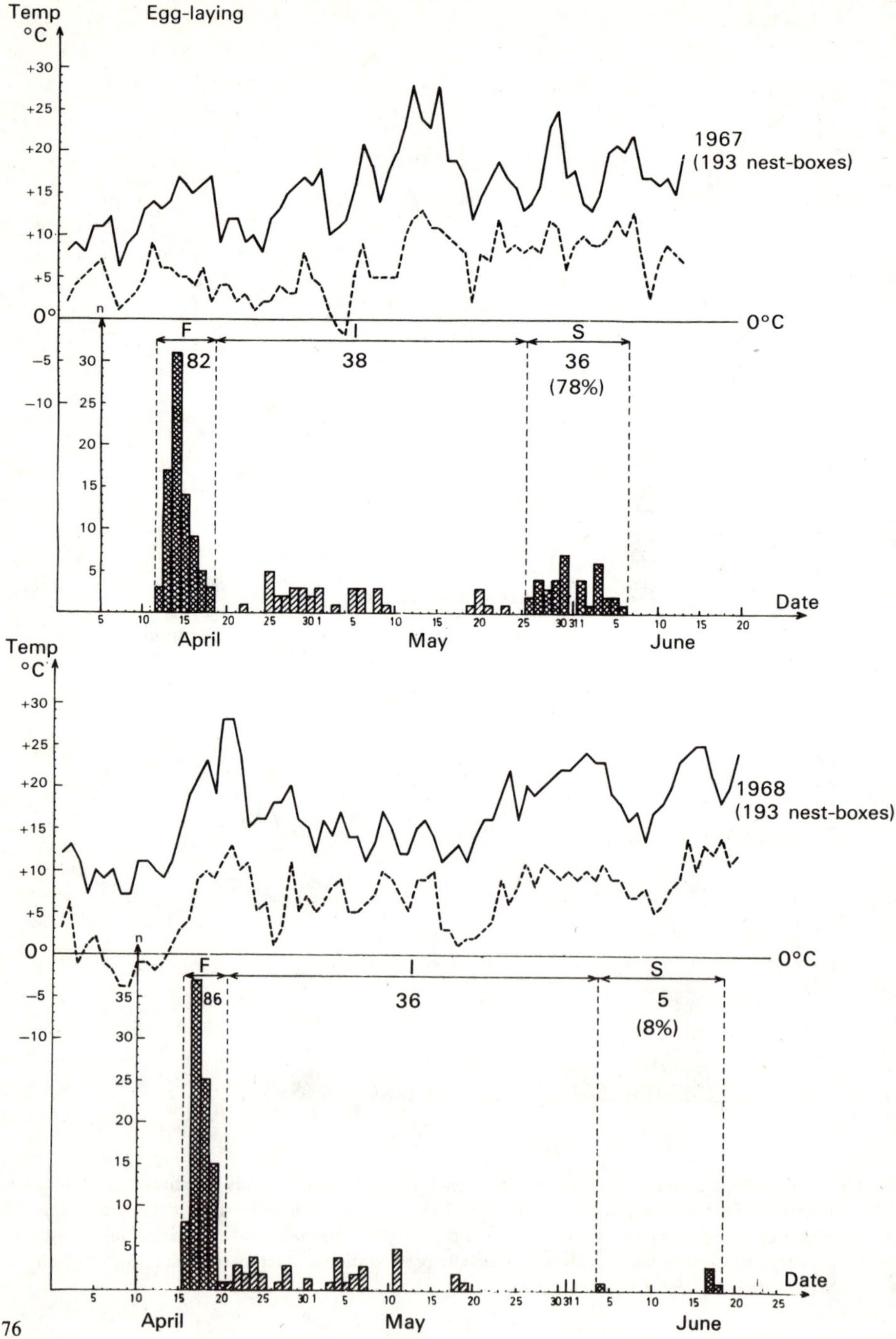
Temp
°C
Egg-laying
+30
+25
+20
+15
+10
+5
0°
−5
−10
n
30
25
20
15
10
5
1967
(193 nest-boxes)
0°C
F
I
S
82
38
36
(78%)
Date
April
May
June
1968
(193 nest-boxes)
35
86
36
5
(8%)

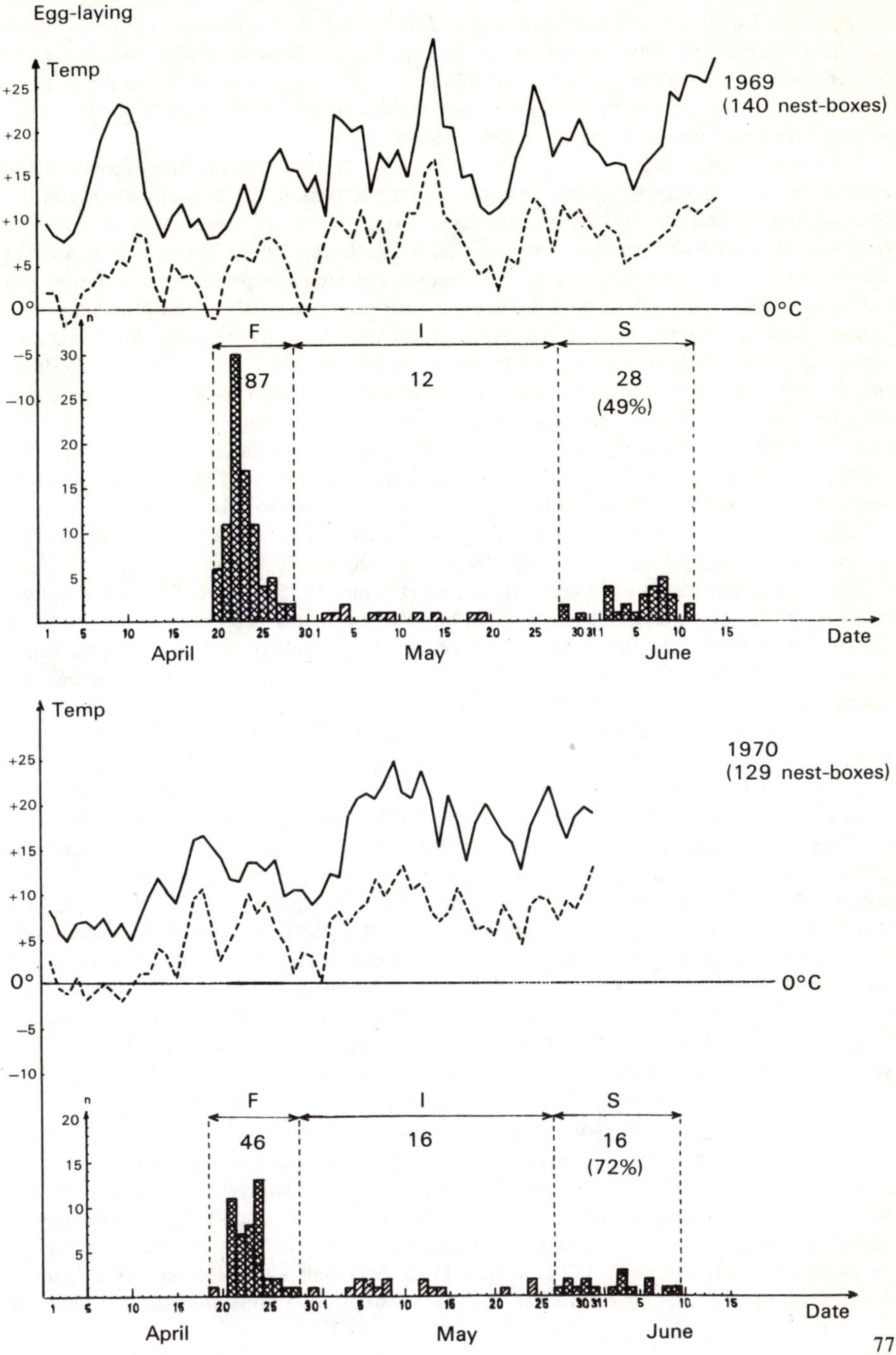
Egg-laying
Temp
+25
+20
+15
+10
+5
0°
−5
−10
1969
(140 nest-boxes)
0°C
n
30
25
20
15
10
5
F
I
S
87
12
28
(49%)
1 5 10 15 20 25 30 1 5 10 15 20 25 30 31 1 5 10 15
Date
April
May
June
Temp
+25
+20
+15
+10
+5
0°
−5
−10
1970
(129 nest-boxes)
0°C
n
20
15
10
5
F
I
S
46
16
16
(72%)
1 5 10 15 20 25 30 1 5 10 15 20 25 30 31 1 5 10 15
Date
April
May
June

synchrony of egg-laying than small ones due to social stimulation, there being no significant correlation between colony size and synchrony. However, large colonies have a tendency to start egg-laying earlier than small ones (>80 first clutches, $\bar{x}$: 16 April, n=3; <50 first clutches, $\bar{x}$: 18 April, n=3). Annual fluctuations in the start of laying, even among large colonies, may be more than one week.

My observations indicate a close relation between nest building, matings (copulations) and the start of egg-laying. One or two days after the peak in the building activity of the female, and some two days before egg-laying, the number of matings, as well as the number of unsuccessful attempts, reaches a peak (Verheyen 1968). We are convinced that the weather conditions during the week preceding egg-laying, especially precipitation and temperature between 0500 h and 1000 h, can considerably delay the start of laying, even if enough protein reserves are present in the flight muscle of the females. Moreover, we observed in 1966 that very bad weather (snow, frost) during the start of laying can delay the ovulation of the first egg of a clutch by several days, so that the synchrony of laying becomes less marked, and can also lead to a decrease in the weight of the eggs (t=3,33; df=155; P<0·001). According to Jones & Ward's (1976) hypothesis, this is probably due to a shortage of protein reserves in the flight muscle. Korpimäki (1978) mentioned that cold and rainy weather can interrupt laying for one, or even two, days. Yearling females, in spite of their late arrival at the breeding area, start laying only a few days after older females (in 1966, $\bar{x}$: 2 days; t=2,01, df=38, P<0·001).

Most of the intermediate clutches are repeats (Dunnet 1955, Kessel 1957); only a small number derive from polygamous males. In our study area the percentage of intermediate clutches with respect to the total number of clutches is relatively high ($\bar{x}$: 23%, range 9–29% over 7 years); this is probably due to desertion of broods after trapping males and females on the nest during the first clutch.

The percentage of second clutches, with respect to the number of first broods with fledged young, fluctuates considerably from one year to another ($\bar{x}$: 50%; range 8–78% over 6 years). Our results are in accordance with those of other investigators in Germany (Creutz 1939, Berndt 1939, Schneider 1957), the Netherlands (Kluyver 1933), Scotland (Dunnet 1955, Anderson 1961) and Switzerland (Schifferli 1957, Wagner 1963), but differ considerably from the findings of Kessel (1953) in North America, who recorded a high and regular 93% of second clutches, and with the results obtained in East Prussia, Latvia and Finland where second clutches are unusual (Schuz 1951, Berthold 1964, Korpimaki 1978). The interval between the fledging of the first clutch and the laying of the first egg of the second clutch in the same nest differs from nest to nest: the mean intervals differ annually and sometimes the differences are highly significant. The synchrony of second broods is less marked than that of first broods and fluctuates considerably from year to year. No significant correlation has been found between the mean start of egg-laying of the first brood and the percentage second clutches.

The abrupt collapse of the gonads (post-nuptial regression phase) at the end of May and the beginning of June, and the development to regeneration phase, is probably due to an interaction of endogenous periodicity with primary exogenous information e.g., photoperiod. Secondary exogenous information, e.g., weather conditions, psychic stimuli and food supply, may considerably retard or accelerate the start of gonad regression (Bissonnette 1931, Bullough 1942, Burger 1953, Marshall 1961, Farner 1967, Schwab 1970, Gwinner 1977). Our results suggest that weather conditions and social stimulation about the period of fledging of the first brood probably play a part in determining the percentage of second clutches. We observed regularly that some females, in spite of successful mating and intensive epigamic display of the male, never started egg-laying. Lack of sufficient protein reserves in the flight muscle of the females, perhaps due to (1) the intensive care of the first brood, (2) insufficient assistance of the male, or (3) a lack of food, may be the proximate factor (Jones & Ward 1976). If egg-laying fails, the pair separates after a few days; the female leaves the nest hole and the unpaired male immediately starts displaying or starts cleaning the nest. If mating does not occur gamosematic display of the male and the frequency of visits to the nest hole decrease gradually until he leaves the breeding area after some 30 days.

Discussion

The starling seems to possess several adaptations that help to maximise reproductive output. Some behaviour patterns differ between the sexes, since male and female investment in the brood is different, but the behaviour of both sexes appears to be adapted to increase breeding success. The selection of suitable nest sites is undertaken initially by the male, a preference being shown for high holes that face east or south-east. The selection of a good site is as important for the male as it is for the female, since such a site will increase the chances of survival of his brood even though he may play relatively little part in its rearing.

Once a site has been acquired, its tenure is aided by various facets of behaviour. For example, males occupy several nest holes early in the spring, or even in the winter, and this allows the older males to take possession of the best sites and to mate with the older females; the male's possession of a suitable cavity appears to be of prime importance in pair formation. The extreme attentiveness of starlings at their nest sites in the morning and from 1600 h until sunset ensures that they are present to chase away potential intruders at the times of day that these competitors are most active in searching for nest holes.

Male starlings devote less time to parental care than females; moreover, a male will never raise a brood deserted by his female, while a female will proceed with parental care after desertion by her mate. During a prolonged absence of his female, the male will start removing nest material, eggs and pulli from the nest and will quickly start pair formation with another female. These observations, together with frequent polygyny, suggest that it is advantageous for males to spend less attention to parental care and more attention to initiating several broods. It appears more advantageous for a female to maintain a suitable nest hole from the start of egg-laying and to raise the brood by herself, since the male plays only a small role in parental care.

Colonial breeding is advantageous from several points of view. It probably promotes early breeding, so that young are in the nest at the best time for raising them and the opportunity for a successful second brood is increased. More broods, whether or not from polygamous males, can be raised simultaneously in nests in close proximity e.g., several holes in the same tree. Colonial breeding also promotes polygyny, so that yearling females, who arrive later than two-year-old or older males and females, can occupy a suitable nest hole during pair formation, mostly with a polygamous male, and can start egg-laying. Yearling males generally arrive somewhat earlier at the nest holes than yearling females but they cannot occupy good nest sites since most of them are already taken by older males; these young males rarely reproduce. We did not study the selective advantages of colonial breeding with respect to the protection of chicks and parents against predators and/or to the exploitation of the food supply.

After the fledging of the first brood, starlings frequently change nest sites and mates. It seems to be advantageous for the male to keep the nest hole after the first brood has fledged, to clean it immediately and to mate as soon as possible with each receptive unpaired female. On the other hand, it should be more advantageous for the female to accompany and feed the fledglings for some hours or days, and then return to the colony and to occupy a suitable nest hole for pair formation. However, in spite of nest building and of successful mating several pairs never lay a second clutch, probably due to a lack of sufficient protein reserves in the flight muscles; in this case, the pair separates after some days; the female leaves the nest hole and the male starts epigamic display.

References

ASKENMO, C. (1977) Some aspects of the reproduction strategy of the Pied Flycatcher, *Ficedula hypoleuca* (Pallas). *Dissertation Ph.D.*, Department of Zoology, Goteborgs Universitet, Sweden.

ANDERSON, A. (1961) The breeding of the Starling in Aberdeenshire. *Scottish Naturalist,* 70, 60–74.

BERNDT, R. (1939) Untersuchungen uber zweite Brut beim Star (*Sturnus vulgaris*) im Braunschweiger Hugelland, *Vogelzug,* 10, 7–16.

BERTHOLD, P. (1964) Uber den Fortpflanzungszyklus sudwestdeutscher Stare (*Sturnus vulgaris* L.) und uber bedingende Faktoren der Brutreife beim Star, *Vogelwarte,* 22, 236–275.

BISSONNETTE, T. H. (1931) Studies on the sexual cycle in birds. IV. Experimental modification of the sexual cycle in males of the European starling (*Sturnus vulgaris*) by changes in the daily period of illumination and of muscular work. *Journal of Experimental Zoology,* 58, 281.

BISSONNETTE, T. H.; ZUJKO, A. J. (1936) Normal progressive changes in the ovary of starlings (*Sturnus vulgaris*) from December to April. *Auk,* 53, 31–50.

BOGUCKI, Z, (1972) Studies on the activity of starlings, *Sturnus vulgaris,* Linnaeus, 1758, in the breeding season. *Acta Zoologica cracoviensia,* XVII, 97–122.

BULLOUGH, W. S. (1942) On the external morphology of the British and Continental races of the starling (*Sturnus vulgaris*), *Ibis,* 6, 225–239.

BULLOUGH, W. S. (1942) The reproductive cycle of the British and Continental races of the starling (*Sturnus vulgaris*), *Philosophical Transactions Royal Society London, ser. B.,* 231, 165–246.

BURGER, J. W. (1949) A review of experimental investigations in seasonal reproduction in birds, *Wilson Bulletin,* 61, 211–230.

BURGER, J. W. (1953) The effect of photic and psychic stimuli on the reproductive cycle of the male starling, *Sturnus vulgaris. Journal of Experimental Zoology,* 124, 127–239.

COLLINS, V. B.; DE VOS, A. (1966) A nesting study of the starling near Guelph, Ontario, *Auk,* 83, 623–636.

CREUTZ, G. (1939) Biologische Beringungsergebnisse bei Staren, *Sturnus vulgaris* L., *Mitteilungen des Vereins Sachsischer Ornithologen,* 6, 18–26.

CROSSNER, K. A. (1977) Natural selection and clutch size in the European starling, *Ecology* 58, 885–892.

DAVIS, D. E. (1959) Territorial rank in starlings. *Animal Behaviour,* VII, 214–221.

DAVIS, D. E. (1963) The role of temperature and light in morning arrival of starlings. *Ecology,* 44, 397–398.

DAVIS, D. E. (1963) The hormonal control of aggressive behaviour. *Proceedings of the XIII International Ornithological Congress,* p. 397–398.

DELVINGT, W. (1960–61) L'étourneau en Belgique, Thèse non publiée. Inst. Agr. de l'Etat, Gembloux.

DUNNET, G. M. (1955) The breeding of the starling, *Sturnus vulgaris,* in relation to its food supply. *Ibis,* 97, 619–662.

ECHARDT, W. R. (1919) Aussereheliche Bruten bei monogamen Vogeln. *Ornithologische Monatsberichte,* 27, 84–85.

ELLIS, C. R. Jr. (1966) Agonistic behavior in the male starling, *Wilson Bulletin,* 78, 208–224.

FARNER, D. S. (1967) The control of avian reproductive cycles. *Proceedings of the XIV International Ornithological Congress* p. 107–133.

FEARE, C. J.; BURNHAM, S. E. (1978) Lack of nest site tenacity and mate fidelity in the starling. *Bird Study,* 25, 189–191.

FREITAG, W. (1936, 1937, 1938) Aus den Leben beringter Stare zur Fortpflanzungszeit I, II, III. *Vogelring* 8, 8–15; 9, 43–49; 11, 1–9.

GALLACHER, H. (1978) De spreeuw. Uitg Het Spectrum, Utrecht/Antwerpen, 205 p.

GWINNER, E. (1975) Die circannuale Perodik der Fortpflanzungsaktivitat beim star (*Sturnus vulgaris*) unter dem Einflub gleich – und andersgeschlechtiger Artgenossen, *Zeitschrift fur Tierpsychologie,* 38, 34–43.

GWINNER, E. (1977) Die circannuale Perodik der Fortpflanzungsaktivitat beim Star (*Sturnus vulgaris*) 16–25.

HAARHAUS, D. (1968) Zum Tagesrhythmus des Staren (*Sturnus vulgaris*) und der Schneeammer (*Plectrophenax nivalis*), *Oecologia,* 1, 176–218.

HAARTMAN VAN, L. V, (1951) Successive polygamy. *Behaviour,* 3, 256–274.

HINDE, R. A. (1967) Aspects of the control of avian reproductive development within the breeding season. *Proceedings of the XIV International Ornithological Congress,* p. 135–153.

JONES, P. J.; WARD, P. (1976) The level of reserve protein as the proximate factor controlling the timing of breeding and clutch-size in the red-billed quelea, *Quelea quelea, Ibis,* 118, 547–574.

KESSEL, B. (1950) Observations of the polygamy and territorial behavior of a male starling, *Sturnus vulgaris, Bird-Banding,* 21, 112–114.

KESSEL, B. (1953) Second broods in the European starling in North America, *Auk,* 70, 479–483.

KESSEL, B. (1957) A study of the breeding biology of the European starling, *Sturnus vulgaris,* in North America. *American Midland Naturalist,* 58, 237–331.

KLUYVER, H. N. (1933) Bijdrage tot de biologie en de ecologie van de spreeuw gedurende zijn voortplantingstijd. *Verslagen en mededelingen Plantenziektenkundige Dienst,* 69, 1–145.

KLUYVER, H. N. (1935) Waarnemingen over de levenswijze van de spreeuw, (*Sturnus vulgaris*). *Ardea,* 24, 133–166.

KORPIMÄKI, E. (1978) Breeding biology of the starling. *Sturnus vulgaris,* in Western Finland. *Ornis Fennica,* 55, 93–104.

LLOYD, J. A. (1965) Seasonal development of the incubation patch in the starling. *Condor,* 67, 67–72.

LLOYD, J. A. (1965) Effects of environmental stimuli on the development of the incubation patch in the European starling (*Sturnus vulgaris*). *Physiological Zoology,* 38, 121–128.

MARSHALL, A. J. (1961) Breeding season and migration. In (*Ed*) A. J. Marshall; *Biology and Comparative Physiology of Birds* 2, 307–339.

MARPLES, B. J. (1932) Starling roosts and flight-lines near Oxford. *British Birds,* 25, 314–318.

MARPLES, B. J. (1934) The winter starling roosts of Great Britain, 1932–1933. *Journal Animal Ecology,* 3, 187–203.

MORLEY, A. (1941) The behaviour of a group of resident British starlings (*Sturnus vulgaris*) from October to March, *Naturalist,* 788, 55–61.

ORIANS, G. H. (1972) The adaptive significance of mating systems in the Icteridae. *Proceedings of the XV International Ornithological Congress,* p. 389–398.

PERRINS, C. M. (1970) The timing of birds breeding seasons. *Ibis,* 112, 242–255.

ROYALL, W. C. Jr. (1966) Breeding of the starling in Central Arizona. *Condor,* 68, 196–205.

SCHIFFERLI, A. (1957) Uber Legebeginn und Zweitbruten beim Star in der Schweiz. *Ornithologische Beobachter,* 54, 1–8.

SCHNEIDER, W. (1957) Ein weiterer Beitrag zur Lebensgeschichte des Stars, *Sturnus vulgaris. Beitrage zur Vogelkunde,* 6, 43–74.

SCHNEIDER, W. (1960) Der Star. Berlin; Neue Brehm-Bucherei, Heft 248.

SCHNEIDER, W. (1972) Sind Populationsstudien der Lebensgeschichte des Stars, *Sturnus vulgaris* L., noch von Bedeutung? *Beitrage zur Vogelkunde,* 18, 310–346.

SCHUZ, E. (1942) Biologische Beobachtungen an Staren in Rossitten. *Vogelzug,* 13, 99–132.

SCHUZ, E. (1943) Brutbiologische Beobachtungen an Staren 1943 in der Vogelwarte Rossitten. *Journal fur Ornithologie,* 91, 338–405.

SCHUZ, E. (1951) Vordringliche Fragen uber die Lebengeschichte des Stars. *Vogelwarte,* 16, 41–44.

SCHWAB, R. G. (1970) Light-induced prolongation of spermatogenesis in the European starling *Sturnus vulgaris. Condor,* 72, 446–470.

SCHWAB, R. G.; LOTT, D. F. (1969) Testis growth and regression in starlings (*Sturnus vulgaris*), as a function of the presence of females, *Journal Experimental Zoology,* 171, 39–42.

TENOVUO, R.; LEMMETYINEN, R. (1970) On the breeding ecology of the starling *Sturnus vulgaris,* in the archipelago of South-Western Finland. *Ornis Fennica,* 47, 159–166.

TROETSCHLER, R. G. (1976) Acorn woodpecker breeding strategy as affected by starling nest-hole competition. *Condor,* 78, 151–165.

VAN DER MUEREN, E. (1977) Dominance hierarchy in a group of caged starlings, *Sturnus vulgaris. Gerfaut,* 67, 267–270.

VERHEYEN, R. F. (1968–69) Bijdrage tot de ethologie van de spreeuw, *Sturnus vulgaris* Linnaeus, gedurende de voortplantingsperiode. Dissertation Ph.D., Rijksuniversiteit Gent, 2 delen.

VERHEYEN, R. F. (1969) Le choix du nichoir chez l'Etourneau (*Sturnus vulgaris*). *Gerfaut,* 59, 239–259.

WAGNER, G. (1963) Zum Brutverlauf beim Star, *Sturnus vulgaris. Ornithologische Beobachter,* 60, 205–213.

WALLRAFF, H. G. (1953) Beobachtungen zur Brutbiologie des Stares in Nurnberg. *Journal fur Ornithologie,* 94, 36–67.

WARD, P. (1977) Fat and protein reserves of starlings. *Annual Report, Institute of Terrestrial Ecology,* Cambridge, England p. 54–56.

WHITE, S. J.; HINDE, R. A. (1968) Temporal relations of brood patch development, nest-building and egg-laying in domesticated canaries, *Journal of Zoology, London,* 155, 145–155.

WYNNE-EDWARDS, V. C. (1929) The behaviour of starlings in winter. *British Birds,* 23, 138–153.

WYNNE-EDWARDS, V. C. (1931) The behaviour of starlings in winter. *British Birds.* 24, 346–353.

YOM-TOV, Y.; DUNNET, G. M.; ANDERSON, A. (1974) Intraspecific nest parasitism in the starling, *Sturnus vulgaris. Ibis,* 116, 87–90.

The starling as a successful forager

J. M. Tinbergen and R. H. Drent
Zoological Laboratory
University of Groningen
Netherlands

Summary

The main findings from a field study on foraging in the starling are described. Mechanisms involved in foraging strategies are emphasised, and our data indicate that decision rules must include detailed site memory and prey quality. The suggestion is put forward that a minimal, or threshold, prey density plays a role in the choice of feeding site. This threshold is defined by the efficiency of prey extraction in relation to the energetic need of the predator. Finally, a broadening of research perspective is urged. The integration of detailed study of the abilities of the predator, such as we have undertaken, with analysis of prey population dynamics is essential if we are to understand how predation contributes to community stability. As a first step we advocate measuring energy flow from prey populations to the predator.

Introduction

There are two main approaches to resolving the man-bird conflict. The most direct is to implement various preventive measures in a systematic fashion, comparing treated and untreated plots. This trial-and-error approach holds the possibility of quick returns but the findings will often lack applicability in other situations. Clearly, the practical approach has its limitations and deserves a complementary integrated study directed towards understanding the underlying relationships between the bird and its man-modified habitat; although often slower and lacking the psychological advantage of generating work aimed at alleviating the problem quickly, results once achieved have a broad application. As an example, the approach taken by Murton and his co-workers in the wood-pigeon (*Columba palumbus*) inquiry (Murton *et al.* 1963, Murton 1968, 1971) provides not only an unequivocal answer to the question whether shooting under a bounty system can depress pigeon numbers sufficiently to yield an economic advantage, but adds besides a well-documented case study on the role of man as a predator. The compensatory nature of the various mortality factors has yielded a basic insight of great generality. It is an advantage for biologists like ourselves, actuated by an academic interest in the basic relations between animals and their food resources, to have this opportunity of entering discussions on the practical implications that our work might have with those engaged in applied research. In our case, where practically all of our research activity is dependent on the close cooperation of the various farmers working the land adjoining our field station, we see this contact as one way of discharging our obligation. Let it be said at the outset that we do not presume to provide ready answers, but hope that at least we are asking the questions in such a way that the broad issues are in fact being attacked.

Our study has been inspired by the conviction that current foraging theory, in its more comprehensive form a theory of resource dependence (MacArthur 1972), suffers from too narrow a data base. While we do not wish to decry the importance of further development of the theory, we advocate a vigorous field programme as a prelude to experimentation under conditions more closely resembling the real world than the situations inspired by the optimality theories currently in vogue. Historically, the subject of prey selection and the interdependence of predator and prey has been tackled in birds mainly by observing parents when feeding their young. In particular, we owe much of our thinking and most of the basic observational facts to Tinbergen and his team (1960) and Royama (1966, 1970) all of whom worked on the great tit (*Parus major*). This fieldwork has been extended by a series of ingenious experiments on tits in captivity (reviewed by Krebs & Cowie 1976. Despite the many advantages that this bird offers the observer, we felt that the quantification of the food supply on the scale required, bearing in mind that the parent must be kept in full view during the foraging bouts and preferably should be amenable to direct observation of the intake rate, could only be studied in a bird in an open environment with a structurally simple feeding habitat. The starling (*Sturnus vulgaris*) combines the salient advantages of the hole-nesting habit with a simple feeding habitat. Not only has the basic biology of the species been thoroughly studied, but techniques for the quantification of the food supply have been worked out (Kluijver 1933, Dunnet 1955, Verheyen 1969, Berthold 1964).

Foraging strategies – definition of the problem

We shall consider the following points: (1) the decision rules governing predation by the starling on animal food in the breeding season, (2) the definition of the minimal density of prey allowing exploitation by the starling, (3) an insight into the consequences of predation for the numbers of predator and prey, which necessitates study of the energy balance of the individual predator. The energy flow models of the population are only a descriptive framework. In this paper we present a point of view; supporting data and details on methods will be presented elsewhere.

Site-specific foraging

The data were collected during the breeding season in a starling colony of 13 nest boxes. The colony is situated on Schiermonnikoog, a small Dutch Waddensea island. One or two nest boxes were equipped with a 16 mm film camera. The camera is triggered, by means of a photocell in the nest opening, by the bird entering the nest and light is supplied by an electronic flash. In this way a complete photographic record of the food brought to the nestlings is obtained. It is possible to estimate the length of the prey species from the records and, with independently measured length-weight relationships, the weight of food given can be determined. From a hide on a platform 4 m high and 40 m from the colony the individual birds were observed on their foraging trips. Landing places and, if possible, the behaviour of the bird were recorded. The data permitted the construction of a map of hunting sites used by the individual parents. These hunting maps can be collated with intake rates deduced from the film records.

Earlier work (Tinbergen 1976) showed that learning plays an important role in foraging. Therefore we chose to concentrate on individual birds. In practice this means that most of our data discussed here were collected on one individual, female 34. In this way we hope to trace the most important factors governing foraging. In a later phase the check on more individuals can be concentrated on the factors identified as relevant by this study.

We first consider hunting in the enclosed polder pasture land in the immediate vicinity of the study colony. In the period of observation the main food item was the leatherjacket, the grub of *Tipula paludosa*. Shifts between fields visited are gradual (Fig. 1) but some caused extrinsically are more sudden. An example is the exploitation of field Z4 on 26 May, coinciding with the disturbance due to mowing (the first spring cut). Similar effects follow the application of fertilisers. As a general rule we are convinced that the changes in foraging location are dictated by differences in food intake rate, since when site choice is

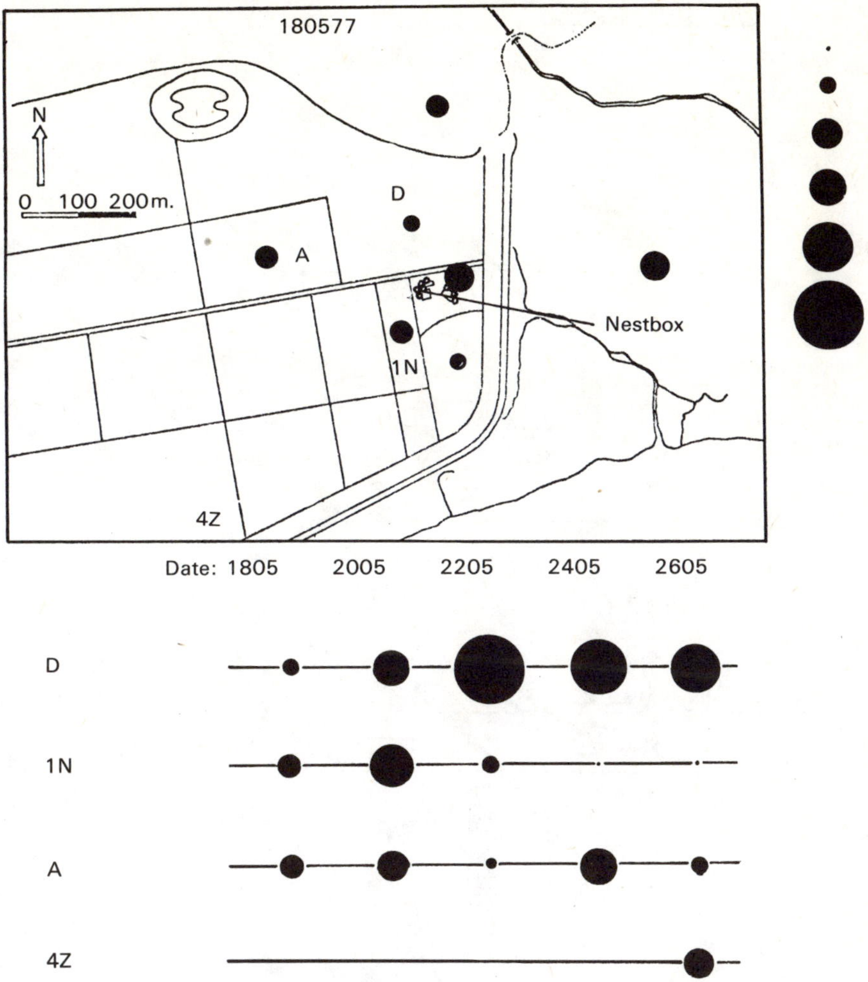

Figure 1. An example of shifts in feeding sites utilised by an intensively watched female (No. 34) while collecting food for the young. Shown is the distribution of foraging sites on various fields (key at the right). Note the sudden appearance of plot Z4 (on that date the farmer mowed the first spring cut with resulting increased availability and enhanced capture rates of leatherjackets). Female 34 was followed continuously from dawn to dusk on alternate days by a team of observers. A more detailed breakdown of foraging sites for the intensive study area is shown in Fig. 2.

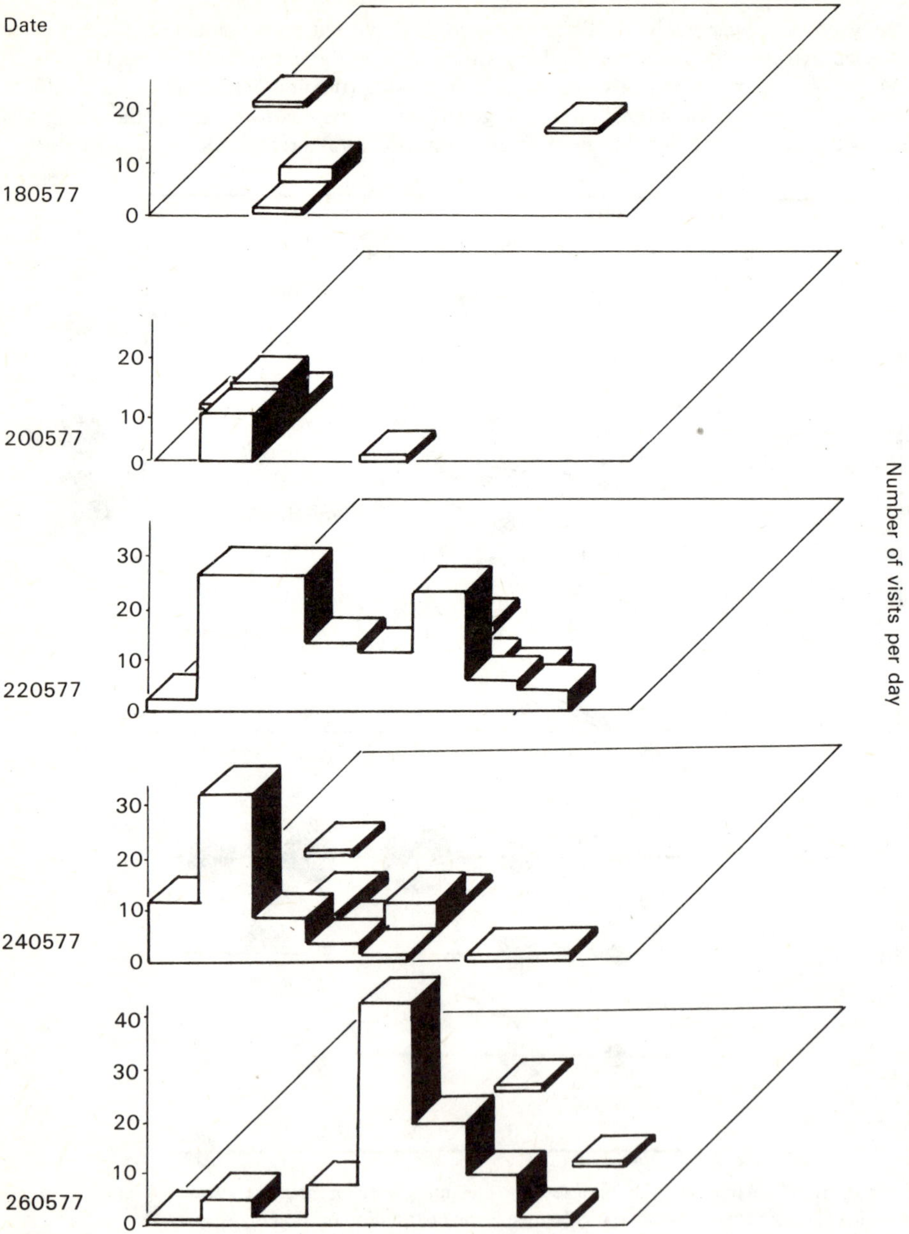

Figure 2. Allocation of foraging sites of female 34 on five dates during the nestling period. The number of landings on 20×20 plots, marked out with pegs, is plotted.

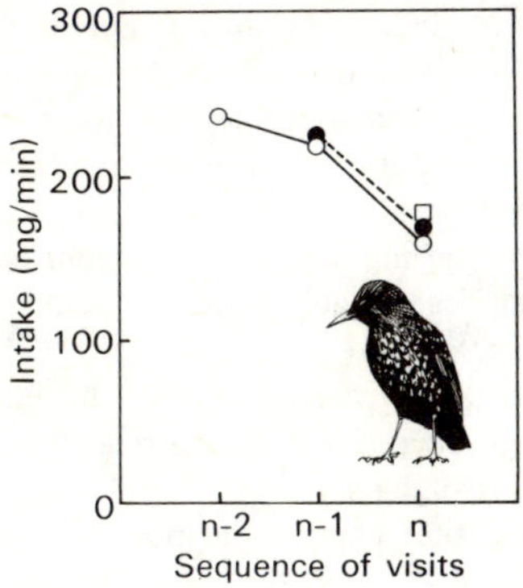

Figure 3. The decision of female 34 whether to return to the same site (defined as the same or an adjacent plot) in relation to the intake rate. If intake rate (leatherjackets primarily) drops below a threshold value of approximately 180 mg/min absent from the nest, the hen shifts to another site. Runs of 1 (■), 2 (●) and 3 (○) visits to a site are shown. The last visit before visiting a new site is n.

viewed within a field on a finer scale (the observation unit is here a block of 20×20 m, as in Fig. 2), it appears that shifts within fields do occur as well. These shifts occur when the intake rate on a block falls below a threshold value (in the case of the leatherjacket this is an intake rate of approximately 180 mg/min, see Fig. 3).

On a fine time-scale, the course of intake rate, as measured during successive visits by the parent landing in the same observational block, shows a consistent pattern. A trend of improvement, as an area is initially exploited, is followed by a tailing off and eventual (temporary) abandonment of the location (Fig. 4). What causes this short-term variation in intake rate? Leatherjackets can emerge from their burrows to feed on the surface during the night, and daily changes in the fine patterns of their distribution can be expected. The depth to which resting leatherjackets descend might make them unavailable to starlings

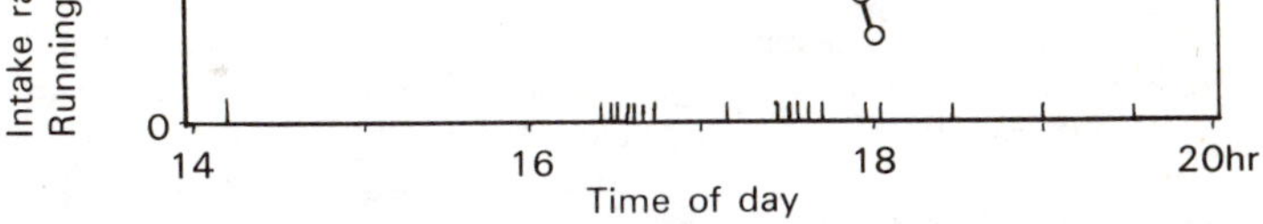

Figure 4. As exploitation of a new area begins the intake rate first improves, then falls away and dwindles to the point of temporary site abandonment. This trend may be due to effects of experience with the local prey distribution (the rise) and subsequent depletion of the stock of available leatherjackets (see text for discussion). Each vertical line marks a visit to the study plot. The intake rate is presented as a running average over four successive visits.

whose effective probing depth is limited to approximately 2·5 cm. This nocturnal movement might render part of the leatherjacket population unavailable to starlings.

We were puzzled by the observed improvement in food intake rate until accurate measurements of the search path of the starlings became available. As shown in Fig. 5, the starling learns where to go, and wastes less time in locating the good feeding areas the more often it returns. In effect, the starling builds up a contour map of intake rates and the pattern of successive landings implies a highly accurate place memory enabling efficient utilisation of local prey concentrations.

Can the decline in intake rate observed subsequently be due to depletion? In direct observations of the birds, paying successive visits to one site only, parameters such as "looking around" and "time spent digging for prey" did not change in the course of a series of visits. This, and the fact that parents did not shift prey types after a bout of

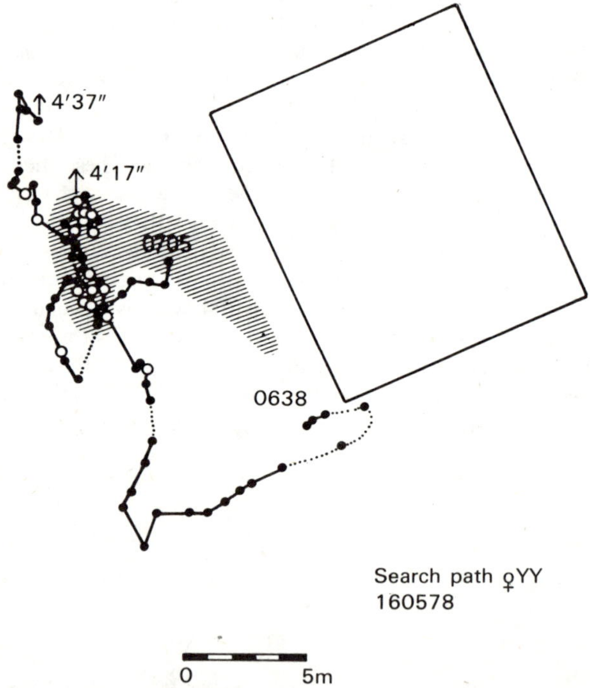

Figure 5. Evidence for success increment as a new area enters exploitation: capture of leatherjackets plotted in relation to the search path in two successive visits of hen yellow-yellow in the egg phase to a short-grass leatherjacket infestation. Key: each dot is a ten second marker, the open circles indicate time markers coinciding with leatherjacket capture. The observers were seated at a distance of 35 m and employed a Zeiss optical rangefinder accurate to within 30 cm at this range. The stippled lines indicate missed time dots, the rectangle represents the field station.

Note that on the first visit of the day in question 3 min elapsed before the first leatherjacket was encountered and captured, whereas on the second visit half an hour later the area giving high intake rates was entered within one minute. For comparison the shaded area indicates the area of high intake rates as deduced from 12 search paths.

exploitation, but continued searching in other places for the same prey species, makes us believe that neither the begging intensity of the young, nor a change in the motivational state of the parents, cause the tailing off of the intake rate. We consider the temporary depletion of the local fóod stock (i.e., exhaustion of the "catchable" segment of the prey population) as the most likely cause of the shift of location. That absolute depletion cannot be a satisfactory explanation is indicated by the estimate that, at most, 1% of the prey actually present is removed during a bout of exploitation. That starlings learn where to go on a very fine scale has also been shown in the laboratory and experiments to elaborate this point are under way.

Optimal solutions to the problem of deciding how to allocate hunting time in the one-prey situation can be derived from a model considering the environment as a number of alternative prey patches with a "no-man's land," involving lost travelling time, in between. Basic to the idea of a patch is the relative uniformity of the prey within such a patch, making it feasible to consider a particular intake rate as characteristic of the patch (Charnov 1973). Our starling data lead us to believe that the real world is more likely to consist of a complex contour map of intake rates of high resolution but of limited duration (i.e., quickly becomes obsolete as prey distribution changes rapidly). Further clarification of (a) the time-scale over which decisions on hunting location are taken, and (b) the area over which the bird generalises, is necessary to achieve a realistic impression of the foraging area as seen through the eyes of the predator. Our present interpretation is that the starling exploits a memory trace of past hunting success comparable in complexity to the display on a radar screen, the images gradually fading away as ever-new impressions are added. This amounts to saying that there is a fine-grained topographic component in the learning ability of the parent starling, making it possible to return to the area with the highest reward rate, or put in another way, to avoid wasting time in areas of low prey intake, which may be a more realistic translation of the problem facing the starling in allocating time.

Choosing between prey types

Kluijver (1933) recorded 313 species of invertebrates in the food collected by parent starlings for their young, but only a few species make up the bulk of the diet. In our study area, eight prey species accounted for 88–90% of the weight of food delivered to the nest, although there were marked variations in the proportions of these items from year to year (Fig. 6). We have seen that the parent chooses between hunting areas on the basis of intake rate, and the simplest criterion for deciding between alternative prey would be likewise to maximise the intake rate (of either calories or nutrients) or perhaps the net rate of gain (intake minus foraging costs). This is not the place to explore this problem in detail, but two general points deserve emphasis. In the first place, to our knowledge no case of choice between different prey types has been satisfactorily explained on the basis of a single criterion. Secondly, the role of the young in influencing which type the parent is liable to bring has been largely ignored so far, and we submit that manipulation of the hunger state of the brood, either in the short-term by deprivation experiments, or in the long-term by manipulation of the brood-size, is an important avenue towards unravelling the selection criteria used in deciding between prey types.

Besides the leatherjacket which we have been considering so far, the caterpillar *Cerapteryx graminis* is an important prey of the starling in our study area. Whereas the

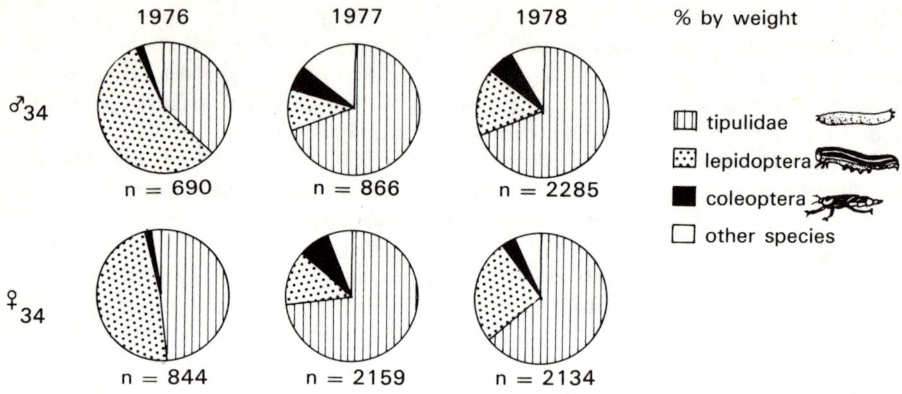

Figure 6. The diet of the starling pair 34 in three subsequent seasons (% composition by weight of the feedings; n = number of feedings to the young on which the figures are based).

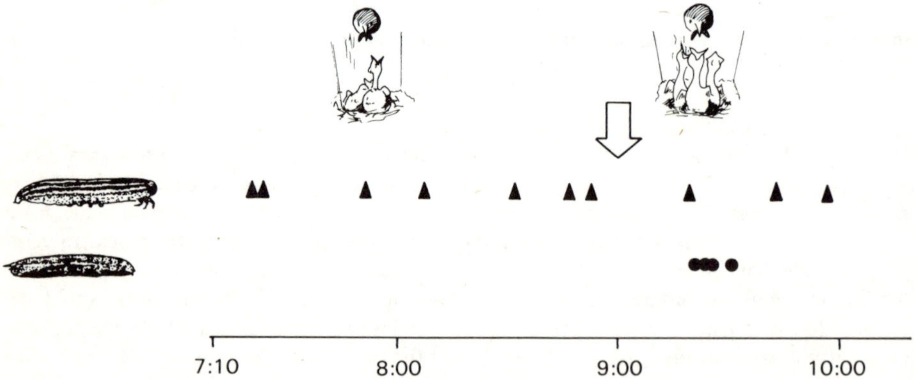

Figure 7. Experiments showing how parent starlings respond to the hunger state of the young. The observations concern a brood of six in a special observation nest (with a removable drawer) which could be observed through a glass partition from a darkened cubicle containing the observer. At 0700 three young were removed from the nest and kept without food until 0900 when they were interchanged with the satiated young. The times the hen arrived with food for the young are shown, a triangle indicating a caterpillar (*Cerapteryx*) brought from the merse, a circle a load of leatherjackets brought from the adjoining pasture land (origin of the prey are determined by simultaneous observation from the tower). Note that the collecting time for *Cerapteryx* is much longer than for leatherjackets.

leatherjacket comes almost exclusively from the enclosed pasturelands near the colony, *Cerapteryx* is found only on the adjoining merseland. Not only is the distance to the hunting areas greater, but the time required to locate a prey item is much longer. *Cerapteryx* is, in short, an "expensive" prey, a luxury item costing two to three times as much as leatherjackets to deliver the same amount of calories to the nest (travel costs included, as computed from Torre-Bueno & Larochelle 1978). Why then does *Cerapteryx* figure so largely in the diet? An experiment provides a clue (depicted in Fig. 7) that *Cerapteryx* is in fact the preferred prey, but that when hard-pressed the parent must compromise between quality and harvesting rate. Given a choice situation at the colony, leatherjackets were delivered to the nest only after the tray with *Cerapteryx* was empty. This response to short-term hunger stimuli (an increase in the proportion of leatherjackets in the diet) is also found in increasing brood-sizes, becoming marked in supernormal broods of seven (Westerterp personal communication). It is revealing that when the weight of prey delivered to the nest was determined by direct weighing (the brood located permanently on the pan of a sensitive balance) it transpired that growth per unit weight of food delivered was lower in broods with a predominant leatherjacket diet, i.e., a monotonous diet of leatherjackets has its penalties, as was discussed by Kluijver (1933). Further work is being directed towards discovering which nutrient components cause this difference in quality, protein being an obvious candidate since *Cerapteryx,* when brought to the nest, is close to the point of pupation and has a high protein content.

Experiments like that depicted here clearly support the idea of an evaluation of the ration by the young, the communication of their need to the parent, and adjustment of parental searching effort within the bounds of the parental time/energy budget. That the nutrient content is an important, perhaps dominant, selection criterion when choosing between prey is not a new idea (Pulliam 1975) but the demonstration that it is mediated by the young in such a proximate fashion is new.

Time patterns of prey capture

As with other birds relying on insect food, the starling must be responsive to changes in prey availability with time of day, and in reaching a decision on which prey type to bring to the nest the time of day is an important restriction. This is illustrated by insect emergence patterns, the insects achieving a degree of synchrony and hence restricting the time of vulnerability to capture. Thus the emergence of *Bibio* occurs within a relatively restricted period each day, the timing of peak emergence varying somewhat depending on day-to-day vagaries in the weather. This is mirrored in the time of maximal delivery to the starling's young (Fig. 8). Is the starling simply tracking the environmental variation in availability or is time a component of the memory system, as was locality? Experiments with food dispensers in the laboratory have demonstrated that the starling can indeed associate feeding at certain sites with the time of day (Daan personal communication) as has long been familiar from work on bees visiting flowers for nectar. Sensitivity to time of day thus plays a role in enabling the starling to utilise evanescent food supplies predictable in the short term. A general discussion on this subject is given in Daan (1980).

We want to mention a further complication of this effect of time on vulnerability of the prey. In foraging studies, it is important to know whether birds concentrate at some particular time of day on a special prey type. Royama (1970) predicted, from his model for great tits, the number of times in succession that the same prey species was brought to the nestlings per day (the "run length" – in this case of bibionid flies), and found that the average observed run length was longer than expected. This led Royama to the conclusion that the birds concentrated their searching in places where bibionid flies occurred. But since vulnerability is dependent on time of day (Fig. 8) at least for starlings, it could very well be that the measured effect is a mere consequence of prey behaviour instead of one of bird behaviour.

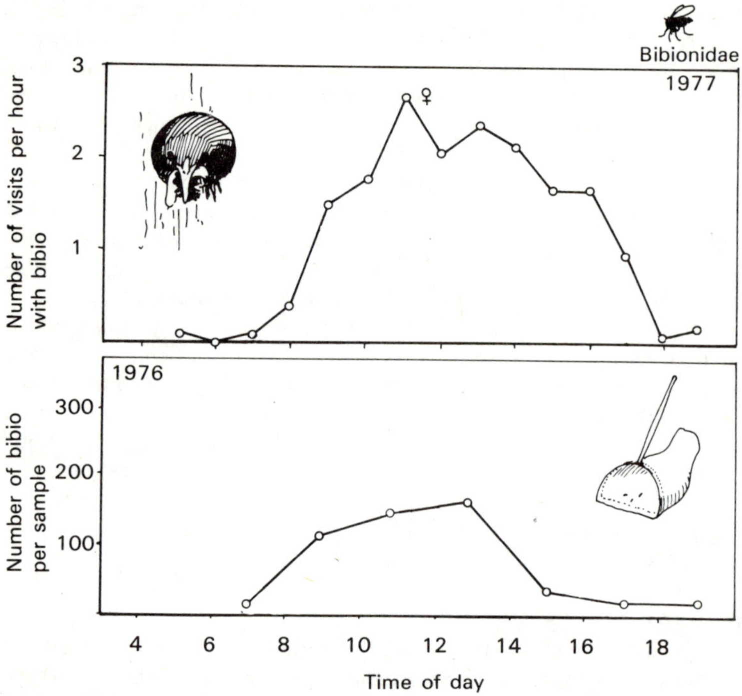

Figure 8. The effect of time of day on the availability and the diet of the young. The data on the availability of the prey were measured with a sweepnet and are shown as an average over two days. The data of the number of feeds with bibionid flies are an average of 10 days and taken from the photographs. The total number of visits changed inversely during the day.

Impact in relation to prey density

Quantification of the impact of predators on their prey is important from the predator's point of view in defining the carrying capacity of the habitat, and from the prey's point of view in determining in which range of densities the predator is liable to influence prey numbers in the long term. Since the starling spends more time in the richer parts of its feeding range, where it enjoys a high intake rate, not only is prey extraction highest in the best areas, but the relative predation rate (proportion of the total prey population removed by the predator) is highest there too (Tinbergen 1976). At moderate prey densities birds may remove sizeable proportions of the insect population but of course the significance that these losses might have for the prey population requires detailed investigation of the population dynamics of the prey; this is rarely included in bird-oriented work (but see East & Pottinger 1975).

In our work we have been particularly interested in defining the prey density at which the starling begins exploitation, i.e., at which prey density does the prey begin to be at risk (the "grazing threshold" as described by Steele 1974). One approach to this problem is to relate mean prey capture rate to the proportion of the prey in the diet. For example, a mean intake rate for leatherjackets in each of five seasons is shown in Fig. 9, with, for comparison, the proportion in which leatherjackets were delivered to the observation nest. Knowing the time budget of the parents (time available for foraging per day) and the food requirement of the brood and of the parent, the required intake rate can be calculated to be 0·75 leatherjackets per minute of foraging time (see arrow in Fig. 9). In 1974, when the leatherjacket was scarce and intake rate fell below this critical level, it was virtually ignored as a food item for the young. This finding presents the possibility that rejection of a prey type may in part depend on an absolute intake threshold. Experiments were

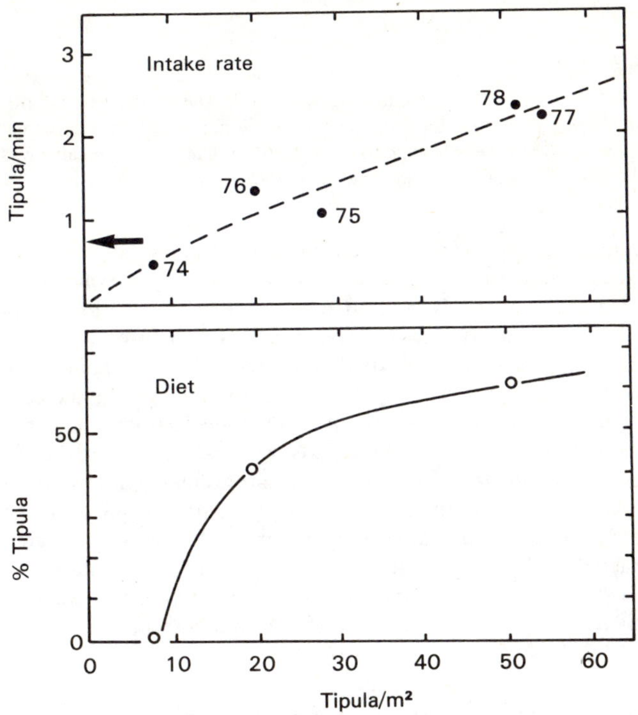

Figure 9. Intensity of utilisation of *Tipula* grubs by starling pair 34 (as percent of trips including *Tipula*) in relation to mean intake rates observed that season (top). There were no complete photographic records for 1975, hence the proportion brought to the nest is not known. Points for 1976, 1977, and 1978 refer to pair 34; for 1974 the value is the mean for two nests intensively watched. Note that the leatherjacket was practically ignored as a food source for the young when the mean intake level fell below the required minimum of 0·75 leatherjackets per minute of foraging time.

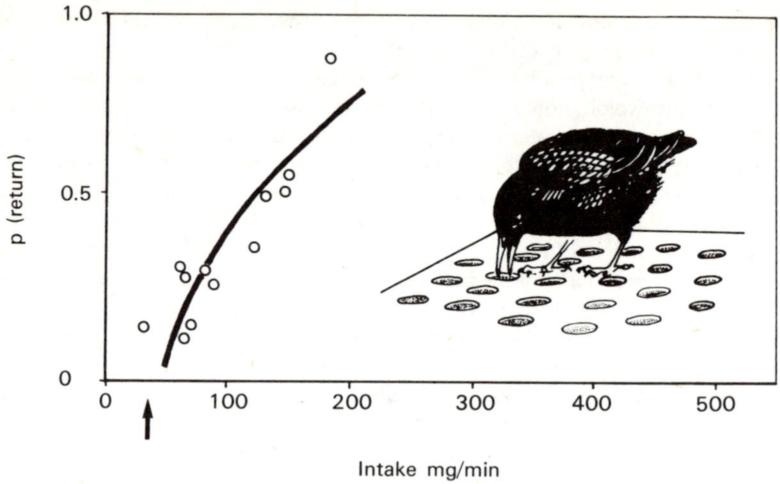

Figure 10. The probability of return to a feeding table stocked with mealworms in competition with the natural food sources during the nestling period. The intake rate on the table could be manipulated by the observer watching from a nearby building. P (return) refers to a time interval of approximately 3 hours (the time interval during which the intake rate at the table was held constant). The arrow indicates the required feeding intake to fulfil needs of the parent + brood.

undertaken during the nestling period by providing a feeding table whose rate of reward (number of pecks to next food item) could be manipulated by the observer watching from a distance. Since the peck rate is virtually constant, the data can be transposed to intake per minute (the data in Fig. 10 refer to mealworms) and, as before, the point at which the table is not visited again during the experimental session of 3–5 hours corresponds to an intake rate that falls below the rate required to provision the young and cover the needs of the parent (indicated by arrow in Fig. 10). A comparable computation can be made on the basis of East & Pottinger's (1975) observations, where starlings abandoned a pasture plot when the density of beetle grubs declined below a threshold density corresponding to the rate of intake barely adequate to cover their requirement (assuming these birds have the same daily food requirement as our starlings in the incubation period, namely 80 kcal per day, and further that the foraging day is 10 hours long). These data do not prove the existence of an absolute threshold for the minimal density for exploitation but indicate that further investigations along these lines should be undertaken.

Discussion

This brief excursion into the natural history of foraging in the starling has emphasised features as yet inadequately incorporated in current theory. The minute topographic scale on which learning operates in the one-prey situation, the quality constraints in deciding which prey to collect for the young, and the existence of a time memory, allowing the predator to concentrate on a specific prey at the time of day during which it is maximally vulnerable, are facets of the starling as a successful forager, which, we believe, will be found to apply to many other predators as well. We finally generalise that there is a minimal prey density below which exploitation by the starling cannot take place. The exact correspondence of this "grazing" threshold with the rate of food intake the bird must sustain to meet its daily requirements, in the three situations for which we have data, has rather baffled us. An influence of the general rate of food collection experienced at the time cannot explain these results. Further work is clearly called for before we can accept this rule for deriving the minimal density attracting exploitation.

The emphasis in our work is on discovering the mechanisms underlying foraging decisions and the hierarchy of these decision rules existing in the individual predator. This type of work may provide a useful framework when we are concerned with problems of birds as pests, but is unlikely to yield insights into the evaluation of the role of birds in regulating the numbers of their prey. Such an undertaking calls for detailed investigation of the population dynamics of the prey, but with the notable exception of a few situations of economic significance there seem to be few such integrated studies where birds are involved. This research vacuum is to be deplored, since from both points of departure our present information is scanty indeed, both as to the relation between bird numbers and the food resource, and the importance of bird predation as part of the density dependent regulation of prey numbers.

Even a cursory glance at the literature shows how widely opinion diverges on the role of predation in maintaining stable interactions between animals and their food supply (neatly summarised in Steele 1974). We feel little sympathy for the deductive approach taken by Hairston, Smith & Slobodkin (1960) and argue instead that an understanding of the relations between animal populations and their resources will come only from study of the processes involved, and not by rejection of hypotheses out of hand.

In short, we would urge those engaged in fundamental research to attempt integrated studies of birds in relation to their food sources. In our opinion a promising approach is to combine estimation of the magnitude of the energy transfer with a detailed knowledge of the behaviour mechanisms in both predator and prey.

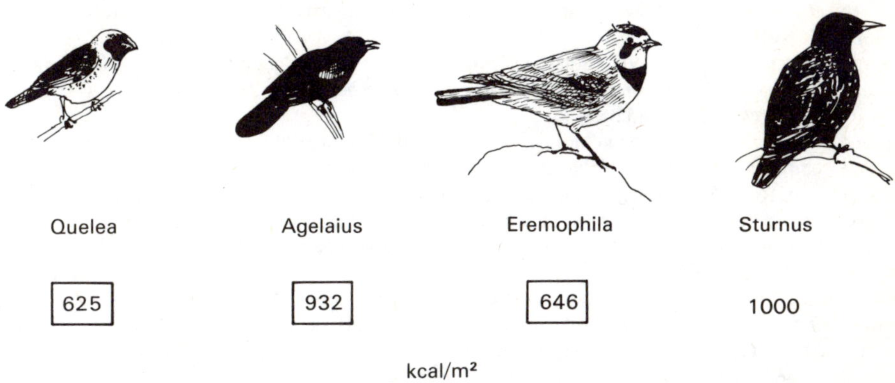

Figure 11. Energy extracted for the nestlings by grassland birds expressed as a total for the colony on the surroundings. Data for red-billed quelea *Quelea quelea*, red-winged blackbird *Agelaius phoeniceus* and horned lark *Ermophila alpestris* are computed from Wiens and Dyer (1977). The value for the starling is an approximation for the enclosed pasture land used by the study colony in 1977.

Such integrated studies are a necessary baseline if we are to understand the role that birds play in enhancing stability of their food base. Current thinking about population regulation (Varley *et al.* 1973, Kowalski 1977) hinges on the consecutive action of a multitude of density dependent mortality agents, basically a succession of predators and parasites operating at different points in time. Recent estimates of the magnitude of prey disappearance attributable to bird predation (Askenmo *et al.* 1977, Dickson *et al.* 1979), in conjunction with the finding that depletion increases with density over a certain range, emerging from the different studies on individual predator species, call attention to the regulatory potential of the bird fauna in its entirety.

As we see it, research should be directed at quantifying the link between food supplies on the one hand and bird populations on the other, with the added incentive that birds as predators may turn out to have important functions in damping oscillations of their prey population. The finding that extraction of energy throughout the breeding process is of the same order of magnitude in diverse grassland based birds (Fig. 11) may serve to justify our optimism.

References

ASKENMO, C.; VON BROMSSEN, A.; EKMAN, J.; JANSSON, C. (1977) Impact of some wintering birds on spider abundance in spruce. *Oikos* 28, 90–94.

BERTHOLD, P. (1964) Ueber den Fortpflanzungszyclus sudwestdeutschen Stare (*Sturnus vulgaris* L.) und über bedingte factoren der Brutreife beim Star. *Vogelwarte* 22, 236–275.

CHARNOV, E. L. (1973) Optimal foraging: some theoretical explorations. Ph.D. Thesis, University of Washington.

DAAN, S. (1980) Adaptive daily strategies in behaviour. In: (*Ed*) J. Aschoff. Handbook of Behavioural Neurobiology; Volume 5; Biological Rhythms.

DICKSON, J. G.; CONNER, R. N.; FLEET, R. R.; KROLL, J. C.; JACKSON, J. A. (*Eds*) The role of insectivorous birds in forest ecosystems. New York; Academic Press.

DUNNET, G. M. (1955) The breeding of the starling (*Sturnus vulgaris* L.) in relation to its food supply. *Ibis* 97, 619–662.

EAST, R.; POTTINGER, R. P. (1975) Starling (*Sturnus vulgaris* L.) predation on grass grub (*Costelytra zealandica* (White), Melolonthinea) populations in Canterbury. *New Zealand Journal of Agricultural Research* 18, 417–452.

HAIRSTON, N. G.; SMITH, F. E.; SLOBODKIN, L. B. (1960) Community structure population control, and competition. *American Naturalist* 94, 421–425.

KLUIJVER, H. N. (1933) Bijdrage tot de biologie en de ecologie van de Spreeuw (*Sturnus vulgaris*) gedurende zijn voortplantingstijd. *Publikaties Plantenziektekundige Dienst Wageningen.*

KOWALSKI, R. (1977) Further elaboration of the Winter Moth population models. *Journal of Animal Ecology* 46, 471–482.

KREBS, J. R.; COWIE, R. J. (1976). Foraging strategies in birds. *Ardea* 64, 98–116.

MACARTHUR, R. H. (1972) Geographical Ecology. New York; Harper and Row.

MURTON, R. K. (1971) The significance of a specific search image in the feeding behaviour in the Wood-pigeon. *Behaviour* 40, 10–42.

MURTON, R. K. (1968) Some predator-prey relationships in bird damage and population control. In: Murton, R. K.; Wright, E. N. (*Eds*) The problems of birds as pests. London and New York; Academic Press.

MURTON, R. K.; ISAACSON, A. J.; WESTWOOD, N. J. (1963) The feeding ecology of the Wood-pigeon. *British Birds* 50, 345–375.

PULLIAM, H. R. (1975) Diet optimisation with nutrient constraints. *American Naturalist* 109, 765–768.

ROYAMA, T. (1966) Factors governing feeding rate, food requirement and brood size of nestling Great Tits (*Parus major L.*) *Ibis* 108, 313–347.

ROYAMA, T. (1970) Factors governing the hunting behaviour and selection of food by the Great Tit (*Parus major* L.). *Journal of Animal Ecology* 39, 618–668.

STEELE, J. H. (1974) The structure of marine ecosystems. Harvard University Press.

TINBERGEN, L. (1960) The natural control of insects in pine woods. I. Factors influencing the intensity of predation in song birds. *Archieves Néerlandaise Zoologique* 13, 265–343.

TINBERGEN, J. M. (1976) How Starlings (*Sturnus vulgaris* L.) apportion their foraging time in a virtual single prey situation on a meadow. *Ardea* 64, 155–170.

TORRE-BUENO, J. R.; LAROCHELLE, J. (1978) The metabolic cost of flight in unrestrained birds. *Journal of Experimental Biology* 75, 223–229.

VARLEY, G. C.; GRADWELL, G. R.; HASSEL, M. P. (1973) Insect population ecology, an analytical approach. Oxford; Blackwell Scientific Publications.

VARLEY, G. C.; GRADWELL, G. R.; HASSEL, M. P. (1973) Insect population ecology, an analytical approach. Oxford; Blackwell Scientific Publications.

VERHEYEN, R. F. (1969) Arrival at the breeding area, maturity for breeding and onset of laying according to age in the starling population (*Sturnus vulgaris* L.) of lower and middle Belgium. *Le Gerfaut* 59, 378–384.

WIENS, J. A.; DYER, M. I. (1977) Assessing the potential impact of granivorous birds in ecosystems. In: Pinowski, J.; Kendeigh, S. C. (*Eds*) Granivorous birds in ecosystems. Cambridge; Cambridge University Press, 205–266.

Discussion

Dunnet: It is very interesting that in the studies of the breeding biology of starlings in many parts of Europe, each place seems to produce a different pattern. For example, in our study in Aberdeenshire, we found that the adult males and females tended not to disappear during the winter. The first year males and females bred later than older birds, but both age categories bred. Sometimes there were pairs of first year birds.

Tahon: Prof. Verheyen described the synchrony of laying of the early clutch. Do all colonies breed at the same time or does each exhibit its own particular synchrony?

Verheyen: Each colony is synchronous but its time of laying may be different from that of other colonies.

Swannack: Has the nutritional quality of the various food items been examined to attempt to explain why the parents may periodically switch from leatherjackets to caterpillars and vice versa?

Drent: There are indications that protein contents vary and that leatherjackets are at their nutritional peak just before pupation. We know that if you feed starlings with leatherjackets and caterpillars weight for weight, growth is less good on the leatherjacket diet. If you provide the birds with a choice of leatherjackets and caterpillars, when the time taken for collection is the same for both types of food item, they select the latter. However, we cannot say, at present, that it is food quality that makes this difference.

Keymer: If starlings select foods that are nutritious for rearing young during the breeding season, is there a danger that they may turn to man-made, balanced foods at this time of year as well as in winter?

Feare: This happened with late brood starlings in Surrey in 1976, when invertebrate food was presumably scarce. Parents fed their chicks on cattle food and the chicks anuses became blocked with the roughage that dried on them. The chicks died and this kind of food appeared totally unsuitable. The high water content of the invertebrates may be important as well as their protein, fat etc. content.

Cowie: In order to switch from leatherjackets to caterpillars a starling has to know exactly where to go. How much time and energy are spent in surveying the area to find where particular food items occur?

Drent: The absolute amount of time a starling spends locating a food source is small, because it profits from the discoveries of others. The graph showing when a starling gave up feeding on a patch showed that it left when its intake rate was still sufficient or just barely sufficient for its requirement. This means that birds do not wander around for long periods with inadequate intake rates. In other words the time taken to decide that an area is poor is very short and therefore the time taken to adequately sample an area is short. I think that is what the evolution of this social system is about: to minimise this sampling time. If a starling can do this by watching what other starlings are doing, it is benefiting from the little bits of sampling time that other birds have put in.

Evans: Can you say anything about social interactions between the birds? Has the behaviour of birds that have poor feeding rates to young been compared with that of those that have high rates of feeding their chicks? In particular, do those with poor feeding rates tend to follow birds with higher feeding rates?

Drent: We do not have any evidence for the Ward and Zahavi hypothesis, but in one year when leatherjackets were particularly abundant, the male of our Box 34 female also had a second female. This male was aggressive and was able to guarantee his females good access to the feeding areas, suggesting that a good forager may have a higher reproductive output than a poorer forager.

Tinbergen: The starlings' intake rate is very high on recently mown grass fields, but the morning after mowing the intake rate drops sharply. When there is a sudden decrease in intake rate like this, there is an increase in the number of birds sitting around. These birds could be looking for others to follow but we have no evidence that they do.

Krebs: Towards the end of Dr Drent's paper he discussed the calculation of minimum energy requirements of starlings and from these predicted whether or not a particular food supply should be exploited. This seems to be an over-simplification because a bird's decision to eat a particular food item will depend on what else is available. If what you are eating is inadequate for your requirements, but nothing else is available, you will still choose to eat it rather than sit there and starve. I would not, therefore, have thought that you could predict the switching off point for tipulids just by looking at tipulids – you would have to know what else was available.

Drent: A bird with complete information would be certain that he should remain feeding at the low intake place, but in fact the bird does not have complete information and is not certain. If he meets a point where he no longer fulfils his energy requirement, the most sensible act is to go elsewhere and invest time in finding an alternative, adequate source. Any time spent remaining in the initial area after it has become non-profitable is time lost.

Feare: Your data on the actual weight of food were obtained from photographs of food items in the nest box, but your field observations must include the food that the bird gets for itself as well as that for its young. Is there any set sequence for a bird to obtain food for itself and then for its nestlings, and does it select different food items for itself and its chicks?

Tinbergen: We only know what the starlings were doing within visual range of the tower. Here we could see leatherjackets being swallowed, and other swallows we assumed to indicate small food items which we knew were present from our sampling of the area. However, we do not know what the birds were doing on the merseland and it will be very important to find out what and how much the parents are collecting there. In front of our observation tower the parent's first interest is to collect food for the young.

Swannack: A recent farming publication claims that tests have shown that starlings, and other wild birds, sometimes carry pathogens, including salmonellae and that an outbreak of salmonellosis at a piggery was probably due to starlings. It seems that this report is inaccurate, which is dangerous when we are considering emotive topics like salmonellosis.

Keymer: I should like to expand on what Dr Feare said about disease. I have not found pathogens in starlings that have been shot, only in those that had died naturally in the roost. The diseases isolated include salmonellae of various serotypes, *Yersinia pseudo-tuberculosis,* erysipelas, gapeworm (*Syngamus trachea*) but, surprisingly, no tuberculosis. We found salmonellae in excreta taken from the roost, but not in excreta taken from cattle and pig houses. We have not looked for viruses because there is little likelihood of finding viral infections in wild passerines except, perhaps, avian pox. Duck virus hepatitis is specific to ducks but starlings could carry it mechanically. The same may apply to Newcastle Disease although starlings are susceptible to it. There is some evidence that starlings can mechanically transmit TGE of pigs. Most of the starling deaths within the roost appear to have been due to starvation and hypothermia rather than disease and it seems that starlings are no more important in spreading disease than gulls or wood-pigeons.

Milne: Since juveniles are responsible for most cherry damage, is there any evidence in years of poor breeding that less damage is done, and could reduction of the yield of young starlings be effective in reducing damage?

Feare: At present we are in no position to be able to limit the output of juvenile starlings, but in any case there seems to be no relation between the "goodness" or "badness" of a starling breeding season and the amount of damage in the same year.

Dunnet: In the dynamiting of a roost presumably the entire site is destroyed.

Tahon: Only the leaves (much laughter).

Slater: Can you indicate the economic cost of killing one bird by this method?

Tahon: That is very difficult. The main component is that of labour, with 100 people involved for three days for each exercise. In addition there are other invisible costs and we have no figure at present.

Dunnet: We have heard of this dramatic technique of attempting to reduce a starling population. Obviously this was a very ineffective way of reducing even a local population for more than a short time. With respect to bird pests in agriculture, are we now in the position where we should no longer consider that we can reduce their numbers and then maintain the low level populations? The number of birds that we can kill seems to be insignificant with respect to population processes.

Jones: Mass killing of red-billed quelea is still recommended for local control in Africa where a colony or roost is close to a vulnerable crop. This can confer protection long enough to allow the crop to be harvested. It is also worthwhile to destroy quelea before the young fledge, because damage is caused mainly by juveniles and these stay around the colony for about six weeks after fledging.

Wright: The philosophy that we (MAFF) currently follow is to identify the problem and then consider the various needs to alleviate the problem. We do not discount killing, and with a sedentary species like the bullfinch, alleviation of the problem may be achieved, but we don't have sufficient data to show this. Our official policy does not include indiscriminate killing, e.g., we no longer kill wood-pigeons simply because they eat crops.

Rogers: Up to about 1972 it was the written policy of the USDA that reduction of the blackbird/starling population was beneficial. Since then, with more data and more enlightened thinking, the policy was changed and at present no more than 10% of the population may be killed, and then only if they pose a documented health hazard or are involved in agricultural damage. We have no data to show that red-winged blackbirds in winter roosts are a problem and there appears to be no case for killing them. Any damage in winter is at feedlots or to sprouting cereals and appears all to be done by starlings. The damage amounts to less than $\frac{1}{2}$% and at present we have no official position on starlings but are concerned mainly with the maintenance and preservation of blackbird populations.

Milne: It would seem unlikely that such control techniques would work with species that have a high reproductive potential and large-scale movements in their annual cycle, but with long-lived, slow breeding species it might be very effective to remove large sections of the population.

Feare: With the longer-term problems of starlings like the eating of cherries over a two month period and of cattle food over a four or five month period, we simply do not know enough about the local movements of birds during the summer and winter seasons to enable us to decide whether local control by lethal means is worthwhile or not. The means of lethal control may have further implications, for example shooting may also have a deterrent potential but where all the birds are killed, as by explosives, subsequent immigrants are unlikely to be deterred from the area, even though the actual roost may be uninhabitable.

Eds. In conclusion, the chairman commented that the papers had amply shown the need for both applied and academic approaches to bird problems on farms. He closed the session by thanking the speakers and the participants in the lively discussion.

BIRD SCARING

Introduction

John R. Krebs
Edward Grey Institute of Field Ornithology
University of Oxford

The first three papers of this section form a cohesive review of behavioural techniques used in bird scaring. They also provide a striking testimony to the close interplay between pure and applied research in this area. Slater and Brémond introduce two ethological concepts, "habituation" and "super-normal stimulus" which are essential as background knowledge in designing effective scarers, and Inglis convincingly demonstrates that knowledge derived from pure research can be applied in a practical way.

The term habituation refers to the simplest kind of learning shown by animals: learning not to respond to a repeated stimulus. Just as people living on a main road become immune to the roar of traffic, so animals come to ignore repeated but irrelevant sounds, sights, smells and other stimuli. Habituation occurs in animals ranging from worms to man whenever a stimulus is repeated without reward or punishment; it seems to be a near universal property of nervous systems. Birds are no exception, and simple scaring devices such as milk bottle tops on a string soon loose their effectiveness. Slater succinctly reviews the voluminous literature on habituation and suggests three ways in which habituation can be countered when using bird scarers. These are: periodic rather than continuous presentation, frequently varying the site of the scarer or the stimulus used, and occasionally reinforcing the stimulus with a real danger such as a gun or a hawk. These are general conclusions based on the nature of habituation, and serve only as guidelines for the design of effective scarers in any particular case.

Animal sense organs are tuned to respond selectively to stimuli which are crucial for survival and reproduction. As a consequence the sense organs may respond in what appears to us to be remarkable and unexpected ways. Many years ago Tinbergen observed that territorial male sticklebacks will direct their aggressive attacks not only to red-bellied rival males, but at also to other red objects including a passing post office van! Ethologists have since created many examples of "super-normal" stimuli: caricutures of a natural stimulus which are even more effective than the real thing in eliciting a response. To mention two famous examples, oystercatchers were seen to try and clamber on to, and incubate outsize eggs, and baby herring gulls preferred to solicit food from a red knitting needle than from a model of their own parent's head. Natures own manipulators, cuckoos, have discovered the same trick and successfully persuade hedge sparrows and other small birds to feed the super-normal yellow gape of a parasitic nestling. Brémond, in his chapter, describes how he has created super-normal bird songs, and goes on to discuss the possibility of super-normal scaring sounds, based perhaps on alarm calls. This intriguing idea is still in its infancy, but in view of the success of visual super-normal stimuli, it seems promising.

One way to find a good scaring device is to screen a very large number of possibilities until, like the legendary chimpanzees at their typewriters, the work produces a scarer as effective as Banquo's ghost. Inglis argues most persuasively for an alternative approach. By using background information from the ornithological, ethological and psychological literature, intelligent guesses can be made as to the likely success of different kinds of stimuli. Inglis uses the term "bio-visual scarer" to refer to methods which take advantage of the fact that certain kinds of stimuli regularly frighten birds in their day to day lives. For example, when woodpigeons take flight they show conspicuous white wing bars. Based on the inference that other pigeons would read the wing bars as a signal of danger, the late R. K. Murton suggested that an effective pigeon scarer might be based on the wing pattern. Wings laid out on the ground have some scaring effect, and work is in progress to design a super-normal wing.

Hawks and other birds of prey are a major natural enemy of many small birds, and while it is not always feasible to fly trained hawks over crops as a deterrent, a hawk-like stimulus should be a good candidate for a scarer. Attempts to use hawk models have not always been very successful, and Inglis suggests that not enough attention has been paid to finding out exactly how birds recognise hawks and whether different features are important for different types of predator.

In the final chapter of this section, Owen raises a profound ecological question. If birds can be effectively deterred from feeding on crops, where will they go? Anyone can observe that garden birds become tamer in harsh winter weather: when they are desperate for food they will take greater risks to get a meal. In an analogous way, the effectiveness of any scaring device will depend on how easily the birds can move to alternative feeding or resting sites. This leads to the idea of establishing refuges for pest species. Dealing with geese as an example, Owen discusses the possible consequences of establishing a network of refuges, and in so doing raises some fundamental questions in population ecology. One consequence of establishing a refuge could be to increase survival of local geese (by providing good feeding sites) and to draw in more geese from outside. Thus the goose population might increase rapidly in size until it spills out of the refuge back into agricultural land! Whether or not this is a realistic picture depends on the nature of the factors which limit goose populations and distribution. This is the kind of problem which has exercised bird population ecologists over the last 30 years, and in addition to any economic advantages which might accrue, the creation of refuges can be seen as a unique opportunity to improve our understanding of the phenomena of carrying capacity and habitat selection.

Bird behaviour and scaring by sounds

P. J. B. Slater
School of Biology
University of Sussex

Summary

This paper starts with a brief account of some of the points which have to be considered when devising techniques for scaring birds by means of sounds. Two particular topics, habituation and auditory localisation, are then discussed in more detail. It is suggested that habituation can best be avoided by minimising the frequency of stimulus presentation and by varying the stimuli used and their location as much as possible. The stimuli used in bird scaring may also vary in effectiveness depending on the ease with which they can be localised. While there is need for more research in this area, current indications are that stimuli which cover a broad frequency range are easier for birds to locate and may therefore be more likely to elicit fleeing.

Introduction

I must start with a disclaimer: I have no expertise in the scaring of birds. Indeed, as an ethologist who spends some of his time working on the vocalisations of wild birds, my efforts tend to be directed to disturbing them as little as possible. However, there is no doubt that many aspects of the behaviour of birds must be taken into account if effective scaring techniques are to be devised, and I would like to discuss some of the relevant issues in this article. I do not intend to provide an extensive review of the bird scaring literature as several of these are already available (e.g. Giban, 1962; Frings & Frings, 1967). Instead, I shall start with a brief survey of some of the points which have to be taken into consideration if time, money and effort are to be invested in bird scaring, and go on to discuss habituation and auditory localisation, two phenomena which, if understood, may help in the development of improved methods.

The simplest bird scaring techniques, making use of scarecrows or loud noises, have an ancient pedigree, but it is only in the last 25 years that scientific work has been devoted to discovering methods which are more effective than these. As far as acoustic devices are concerned, a number of possibilities have been tested. The simplest of these rely on the aversiveness of loud noises, such as those produced by fireworks, but these tend to be relatively ineffective for long-term use and to have nuisance value for humans and other animals. Potentially at least, methods which rely on reproducing the calls of birds themselves have much more to offer. There are three main advantages (Frings & Frings, 1967). First, such sounds may be effective at considerably lower intensities, making them less costly to produce once the necessary equipment has been purchased and also less annoying to humans. Second, it has been argued that habituation to these natural calls will be slower because animals are adapted to show particular responsiveness to them. Some workers using such sounds have even failed to find habituation, a point which will be taken up further below. Thirdly, such sounds vary in the extent to which their effects are specific to the particular species which produces them. It is therefore possible to choose calls which will influence one species but not others or ones which have a more general effect.

Many of the natural sounds of animals elicit approach or withdrawal from other individuals, and calls having either of these influences may be useful in ridding an area of unwanted birds. In some species, feeding or assembly calls may attract individuals and could, in theory, be used to move them from an area where they are undesirable. But it is doubtful if such methods would be of much use with individuals which have already found a good supply of food. Scaring birds away with the calls of their own species, or sometimes with those of predators, is the option which has usually been selected. Two classes of calls are particularly relevant here. Birds which are maltreated or restrained often produce piercing shrieks known as distress calls, which are loud and cover a wide range of frequencies. Alternatively, most species have alarm calls which are more restricted in frequency and are produced in response to sighting a predator. Field tests suggest that the latter type of call is, in general, more effective, but the results on both show considerable variability with both local and species differences in responsiveness. Even where responses are found, species may differ in the exact form which these take. For example, Brough (1968) reports that distress calls played to starlings lead them to fly straight off, while gulls are attracted to the speaker and circle round it before dispersing.

Many different factors are likely to contribute to the varying effectiveness of different signals. One, which is of particular importance from the economic point of view, is the quality of the equipment used in sound reproduction. To be effective some calls may need high fidelity reproduction while the response to others may be present despite the distortions introduced by cheap equipment. The group size in which animals are feeding may also influence their tendency to flee. Reports on this vary, but some suggest that large flocks are more easily scared than small ones. While the threat from a predator to an individual is usually less in a flock, and the response of grouped birds might therefore be expected to be lower, larger flocks are more likely to contain individuals close to the loudspeakers, new to the area or with a lower threshold for scaring. One or two birds which respond strongly because they are in one of these categories may get the whole group moving.

Another factor of importance is the availability of alternative places to go. Not all work on bird scaring has concentrated on removing them from crops. The tendency of some bird species to rest on airfields and to roost on public buildings has also had undesirable consequences which have led to efforts to repel them. The problems here are rather different. Populations of birds are not in general limited by the availability of roosting or resting places so that moving them to other, less undesirable sites is a reasonable goal. In the case of scaring birds from feeding places there is one advantage and this is that a permanent shift out of the area is not required but only one that lasts for the period during which the crop is vulnerable. On the other hand if, as is often the case at least at some seasons, the population of birds is close to the limits of its food supply, then it is unlikely that scaring techniques will provide a solution to the problem. Hungry animals will take great risks to get food: one cannot expect them to starve when it is available, albeit accompanied by an intermittent alarm call, on the other side of a fence. The likely effectiveness of scaring techniques does therefore depend on whether the population in question could find sufficient food elsewhere without making depredations on human crop plants. Of great importance too, though perhaps not to the income of the individual installing scaring equipment, is how mobile the population is. To scare birds from one farm merely to double their numbers on another will hardly lead to good relations between neighbours or an improvement in the national economy.

These considerations suggest that scaring techniques are no panacea, although they may, in the short term, provide a solution to particular problems. Amongst these are the occurrence of abnormally high concentrations of individuals in a particular place or the attraction of birds away from a food which, while perfectly adequate, is less palatable or easily gathered that that provided by crops. I would now like to consider in rather more detail the problem of habituation as this is obviously crucial to the effectiveness of scaring methods.

Habituation

The phenomenon

Repeated presentation of the same stimulus to an animal tends to lead to a decline in responding. There are several possible reasons for this, the simplest of which are that the animal's senses become adapted so that it no longer perceives the stimulus, or perceives it less strongly, or alternatively that the muscles involved in the response become exhausted. These possibilities, sensory adaptation and muscular fatigue, are most likely to occur with stimuli which are repeated very frequently. In many cases of response waning both these mechanisms can be discounted. If, for example, the animals can still use the same muscles for other responses, then muscular fatigue cannot have taken place. Likewise, if the animal still responds to the stimulus, but in a different way, then it is clearly still able to perceive it. Tests such as these suggest that many cases of declining responsiveness cannot be explained so simply and must be attributed to changes taking place within the central nervous system. Where changes of this sort are more or less specific to a particular stimulus and relatively long term they are usually classified as examples of habituation.

Rather little work has been carried out on the habituation of birds to auditory stimuli and, as mentioned earlier, reports differ on the extent to which the phenomenon is of importance in the practical application of scaring methods. Habituation is, however, a widespread phenomenon, and certain broad generalisations can be made about it (Thompson & Spencer, 1966). Detailed studies show that exceptions to these exist in particular cases (see Hinde, 1970), but some features are of sufficient generality to be worth discussing as a basis for deciding how habituation is best minimised. I shall first list these characteristics and then illustrate the phenomenon and some of its hidden complexities by reference to two particular series of experiments. The following features of habituation are amongst those given by Thompson & Spencer (1966):

1. *Responsiveness decreases with number of trials.* Both the strength of the response and the probability that it is elicited declines with the number of stimulus presentations. For example, a startling stimulus, when first applied, may cause a bird to flee, later fleeing may become less frequent although the animal still orients to the stimulus when it appears, and finally even orientation may disappear after many tests.

2. *The response recovers with time.* If stimulus presentation is discontinued and then restarted later, the recovery of the response depends on the time that has elapsed without testing. Complete recovery may, however, never be achieved or take a very long time.

3. *With repeated series of trials habituation becomes progressively more rapid.* Thus in each successive series, with time for recovery in between them, the number of presentations required for habituation to be achieved declines.

4. *Weaker or more frequently presented stimuli lead to more rapid habituation.* Weaker stimuli tend to elicit a less strong response in the first place and the response declines to zero on fewer trials. The effect of frequency of stimulation is, at least partly, because at high rates there is little time for recovery between trials.

5. *The later effects of habituation are often increased by continued stimulation after responding has ceased.* Recovery, for example, is reduced or absent if stimulus presentations are continued during the period after responding has ceased.

6. *Habituation exhibits stimulus generalisation.* Animals which have habituated to a particular stimulus will also show less response to similar stimuli, the greater the similarity the more they will generalise.

7. *A different stimulus may give dishabituation.* The response may show partial recovery if the stimulus is changed. The effect is strongest if the new stimulus is very different from the previous one but changes in location or loudness may also operate in this way.

The widespread nature of this phenomenon amongst animals may be illustrated by two examples. A particularly clear one comes from the work of Wolda (1961) on the water boatman (*Notonecta glauca*). These animals turn towards disturbances on the surface of the water and Wolda studied changes in this response to touching of the surface with a fine wire 2 cm from the animal when this was repeated at 5 s intervals. Both the probability of responding and the intensity of the responses which were made, measured as the angle through which the animal turned, declined to zero over a few hundred trials. Recovery only took place if stimulation was discontinued after habituation had occurred and it was a slow process: after 24 h initial responsiveness was high once more but the decline was more rapid than in a fresh animal. The decrease in responding was, however, specific to a particular location of the stimulus: an animal which had been given 300 trials with the stimulus on one side of its body would respond just as strongly again when these were followed by 300 on the other. Indeed, recovery of the response on one side was actually enhanced by stimulation on the other during the rest period.

Turning to a species more relevant to our present purpose, Hinde (1954, 1960) carried out an extensive series of experiments on the mobbing response of chaffinches (*Fringilla coelebs*). The complexity of the findings do not permit easy summary, but a few points are worth making. A chaffinch, when first confronted with a stuffed owl, flies around it and makes "chink" calls. The amount of calling declines over 30 min or so and then recovers during the following two hours if the owl is removed: recovery is not, however, complete and even 24 h later the animal will only produce about half the number of calls that it did on the first test. The longer the initial presentation the greater was the decline on retesting and, in general, the longer the interval, the more recovery was found. However, responsiveness showed marked fluctuations with time which defied interpretation in terms of a single process. For example, when the owl was first introduced, the response was not maximal immediately but rose over several minutes. Exhaustion of mobbing to an owl also decreased that to another predator, such as a model dog or snake, when this was presented immediately afterwards. But if the dog, which was a weak stimulus, was presented after some recovery it was actually mobbed more the longer the period that the chaffinch had first been tested with the stuffed owl. To account for all of his findings Hinde suggests that some of the changes are specific to a particular stimulus while others affect the mobbing response as a whole. Furthermore, the changes seem to result from a mixture of short, medium and long term processes, some of them increasing and some decreasing responsiveness.

These results argue against hasty generalisation and show that a blanket term such as habituation is not appropriate as a full explanation of all examples of changing responsiveness. It may be combined with other processes and, in some situations, these may be of sufficient strength to override its effects.

Habituation and extinction

Habituation bears similarities to the phenomenon of extinction, which has been extensively studied by psychologists (Kling & Stevenson, 1970). In this case an animal is trained to respond to a particular stimulus by being rewarded when it does so or punished when it does not, and extinction is the process whereby it ceases to respond when the reward or punishment is withdrawn. The two processes are more than just analogous, for many of the responses used in studying habituation may have been built up through learning during the lifetime of the individual even if not by experimental training.

Conditioning experiments are amongst those which have shown that birds find certain calls aversive. Male chaffinches will learn to land on a particular perch if this switches off a tape playing mobbing calls (Thompson, 1969). Likewise, some jackdaws (*Corvus monedula*) have been trained to peck a key to avoid hearing distress calls of their own species (Morgan & Howse, 1973, 1974). Such calls are probably naturally aversive to individuals which have never heard them before but, in the normal life of the animal, the response to them may be maintained at a high level because of their association with alarming circumstances such as fighting and the appearance of predators.

Experiments with rats and pigeons have shown that extinction is slower in some circumstances than it is in others. In general, the more similar the extinction conditions to the training ones, the slower will be the disappearance of the response. An animal which has been trained to press a bar for food reward which is delivered every time the bar is pressed (continuous reinforcement) ceases to respond rapidly when the reward is witheld. But if ten presses are required for each item of food (a fixed-ratio schedule of reinforcement), bar pressing persists for longer when the food ceases to be available.

There is an interesting possible parallel between experiments such as these and the responses of animals to alarm calls. Amongst the circumstances in which alarm calls are normally produced is that in which an individual spots a predator. Other animals then flee without necessarily seeing the predator themselves. This is therefore equivalent to a training schedule in which punishment is only delivered intermittently: the sort of schedule on which extinction is very slow. This may be amongst the reasons why birds persist in responding to alarm calls played from loudspeakers for longer than they do to other noises. Furthermore, as they will still hear some alarm calls which are produced by other members of their flock and associated with real danger, the scaring procedure is probably equivalent to raising the ratio of reinforcement (e.g. making punishment less frequent) rather than to extinction. Rydén (1978b) has shown that nestling great tits (*Parus major*) can be induced to show less withdrawal in response to the "seeet" alarm call of their species by its presentation in association with feeding. Thus conditioning can certainly modify the effect of this call and it is possible that similar processes also have an influence on maintaining the response to it in the natural environment. However, these points are somewhat speculative: there is no doubt that birds cease to respond to alarm signals comparatively slowly (see, for example, Rydén, 1978a), and there are good reasons why this should be the case without invoking learning theory to provide an explanation.

Habituation and scaring

How should scaring methods be devised so as to minimise the possibility of habituation? From the characteristics of this process discussed above, certain guidelines are obvious. The following points may be worth stressing:

1. Stimuli should be presented as infrequently as possible. The less frequent the stimulus, the more time is allowed for recovery between responses. The timing of stimuli should therefore be based on the period it takes for sizeable flocks to gather again after they have been scared away. This will minimise the possibility that birds which have been scared into nearby bushes continue to habituate even though they are not feeding. The number of stimuli necessary may also be reduced if presentations are restricted to the times of day when birds feed most: in many species there is a peak of feeding in the early morning and evening.

2. Stimuli should be varied as much as possible. Variation can take a number of different forms. Changes in volume may not help much as they may simply make the stimulus less effective on some occasions than on others. It may be more productive to vary the type of signal, incorporating various different calls or other noises into the schedule, so that each one is used less frequently. The location from which the call comes, or appears to come, can also be varied by moving it between loudspeakers or varying the balance between them. Finally, the temporal pattern can be varied to make it as unpredictable as possible so that birds cannot anticipate the occurrence of a stimulus.

3. Occasional reinforcements? To explore the possible role of the learning processes discussed above, the occasional accompaniment of an alarm call by genuine danger, such as gunshot or the sight of a hawk or human, may help to maintain the response through conditioning.

All these tactics may give more effective and long-lasting scaring but, of course, their usefulness as well as their practical and economic feasibility are likely to vary from species to species and situation to situation. There is a need for more carefully controlled tests on this topic.

Auditory localisation

Is localisation important?

Animals do not always respond to frightening stimuli by fleeing. An alternative response is to become immobile (freeze) for a period and then to recommence activity. From the bird scaring point of view this is clearly a less desirable result. The reasons why one or other of these different reactions occurs in a particular situation have not been studied as extensively as they should have been. There are, however, pointers. Fentress (1968) found that voles which are already active tend to flee, while freezing is more common when they are resting. This is probably adaptive because animals which are active, as for example when they are feeding, are more likely to be far from cover and will thus be more vulnerable. Another factor is probably the extent to which the stimulus can be localised, and there is some evidence that diffuse stimuli lead to freezing whereas those that can be located are more likely to give fleeing (see Archer, 1976). This again seems adaptive, as fleeing in the presence of a predator which has not been located may simply lead the animal into its jaws.

If this argument is correct, efforts should be made to make frightening stimuli used in bird scaring as easy to localise as possible, even though it may be necessary to move them around between trials to avoid the increased habituation which is likely to result. The relevance of this point to alarm signals is less easy to assess. It is, of course, true that the location of an alarm call does not indicate that of the predator, but merely that of a bird which has seen it. Nevertheless, the predator is likely to be in the same area as that from which it comes. An additional problem is that some bird alarm calls are thought to be specifically adapted to make localisation difficult (Marler, 1955): these are the thin and high-pitched "seeet" calls which occur in essentially similar form in many species. These calls are also referred to as "hawk alarm calls", although they are certainly also elicited by dogs, cats and humans during the breeding season. Much has been written about both their structure and their function, yet the evidence on how they influence other individuals is largely anecdotal. On the basis of his observations, Marler (1956) suggests that they make other individuals flee to cover and freeze there, but without experimental data it is hard to be sure that these responses arise from the call rather than the presence of the predator. The same call can occur during courtship without evoking fleeing, and similar calls sometimes accompany threat postures during fighting (Andrew, 1957).

To my knowledge, no one has systematically tested the influence of these calls on other individuals. However, the work of Leger & Owings (1978) on Californian ground squirrels (*Spermophilus beecheyi*) is of interest in this context. These animals have several alarm calls, most of which cover a broad frequency range and are probably easy to locate. One, however, is a bird-like whistle. When female squirrels were played tape-recordings of the calls, most of them evoked an upright posture and scanning combined with persistent disruption of feeding. But the animals delayed showing the upright posture for about 1 min after the whistle was played and also reverted to feeding more rapidly after tests with this stimulus. Leger and Owings suggest that this response is adaptive because the whistle, like the "seeet" call of small birds, is given largely in response to aerial predators such as hawks. By remaining still the animals are likely to avoid detection but, because hawks move rapidly through an area, they need not do so for long. The other calls indicate the presence of ground predators: scanning enables them to be located and, as they stay around for longer, it must persist until this is achieved.

The effectiveness of alarm calls which are difficult to locate in giving escape responses is therefore open to question, and further experiments are badly needed both to discover the reactions of birds to different calls and whether or not the direction of flight is influenced by the location of the call. If the location of the sound is an important variable, then preference should be given to the use of stimuli which birds find it easy to localise.

How do birds localise sounds?

There are two aspects to localisation: the assessment of direction and the assessment of distance. The latter may be achieved in several ways (Moore, 1977), two of which are probably particularly important in the case of naturally occurring signals. Unless the output volume varies, amplitude alone may give a measure of how far away a sound is.

Although the habitat may introduce some complexities (Morton, 1975), the amplitude of sound generally declines by 6 dB for each doubling of distance, so that the distance of a sound can be estimated by a comparison of its loudness with that which it is known to have at source. Alternatively, again with a sound the source characteristics of which are known, distance can be assessed from the fact that high frequencies are attenuated more than low ones, the higher frequencies tending to bounce off objects rather than travel round them. Sounds with a disproportionate low frequency component are thus likely to be further away. This may be one reason why low fidelity recordings, which lack the higher frequencies, are sometimes found to be less frightening to birds (Morgan & Howse, 1973). From the point of view of scaring, the most effective results may therefore be obtained when both the frequency range covered by the signal and its amplitude suggest a nearby origin.

Assessing the direction of a sound source is, in theory, a rather more difficult task. Differences between the ears in the time of arrival of a brief signal, or of successive waves of a long one, could provide cues. Likewise, intensity will vary between the ears if the sound comes from the side, especially with high frequency sounds which bounce off the head leaving a "sound shadow" on the far side. These differences are known to be important in mammals (Mills, 1972) and Marler (1955) argued that, if they were also used by birds, sounds covering a broad frequency range would be easy to locate, whereas those like the "seeet" alarm call which were thin and high pitched would present difficulties. There are, however, reasons why none of these methods would be very useful for small birds. With a head 2 cm wide an appreciable sound shadow would only be left by noises containing frequencies much higher than those occupied by bird calls (about 17 kHz and upwards). The maximum time difference between the ears, when the sound was at right angles to the head, would also be so small (about 58 μs for a head 2 cm wide) as to make it very unlikely that this cue could be used. Added to this problem is the fact that for all directions other than this time of arrival differences would be ambiguous, a sound at 45° to the beak, for example, giving the same difference as one at 135°.

How then do birds localise sounds, without the aid of the pinnae which assist this process in mammals? Two approaches have been used to try and resolve this problem. I have conducted experiments on the head movements shown by zebra finches (*Taeniopygia guttata*) when orientating to a series of clicks. The results, though still preliminary, indicate that their first response is to move the head so that it is at right angles to the loudspeaker. This suggests that they are maximising some difference between the ears rather then minimising it. But what difference is it that they are using? Recent experiments by Coles *et al.* (in press) on quail (*Coturnix coturnix*) have produced the answer. There are air spaces within the heads of birds which connect together the two middle ears. As a result sounds reach each ear both from the outside and through the head from the other ear. The cochlear microphonic potential (a measure of the extent to which the cochlea is stimulated) varies markedly with the angle of the incident sound, but only if the other ear is unblocked so that sound can travel through the head. Strong differences in the potential were found between the two ears, considerably greater than would have been possible if the ears had been stimulated only by sounds reaching them from one side. The magnitude of these differences varied both with the sound frequency and between individual animals, but some general rules emerged. The differences were least with tones of arond 820 Hz and 3·5 kHz suggesting that quail would have difficulty localising sounds of these frequencies. The differences which existed outside these regions were also frequency dependent, however. At lower frequencies there was a tendency to find that a speaker within a single area to one side of the bird gave strong differences between the ears. Outside this area differences were very small for all other angles suggesting that accurate localisation could be best achieved by maximising the difference between the ears rather than minimising it. This fits in with my observations on what zebra finches do when localising sources of sound. In the higher frequency ranges, above 5 kHz, the difference between the ears tended to be sharply tuned to particular directions, but was likely to have more than one preferred direction. This suggests that the bird might have difficulty localising sound in these ranges due to ambiguities, more than one direction providing it with similar cues.

The extent to which these results on quail can be generalised to other species remains uncertain, but is seems likely that the principles will be the same even if the details differ. Thus the exact frequencies at which localisation is difficult may vary from species to species, depending on such factors as head size, but the difficulty of localising high frequency sounds due to ambiguities may be widespread. On the other hand, Shalter (1978) has argued that some predators can locate the "seeet" call, although the measures he used were not very precise. Thus, while the mechanism used by birds in localising sounds has been elucidated, there is urgent need for more work to discover just how well it enables birds to localise different stimuli.

Localisation and scaring

This section has drawn attention to the possibility that frightening sounds, be they loud noises or alarm calls, may be more effective in eliciting fleeing if they can be easily localised and appear to come from close by. The nearness of the source of an alarm call could be assessed by its amplitude or by the proportion of high frequency components that it contains: scaring equipment may therefore be more effective if it can achieve both good volume and high quality reproduction. The assessment of direction is likely to be facilitated by the use of sounds which cover a broad frequency range: high-pitched sounds may lead to ambiguities and pure tones at lower frequencies may fail to produce adequate differences between the ears for localisation to be possible. Beyond these general points the message is a clear one: more controlled experimental work is needed on the reactions of birds to sounds of different characteristics and, in particular, to alarm calls, if the most effective stimuli for scaring are to be discovered.

References

ANDREW, R. J. (1957) The aggressive and courtship behaviour of certain Emberizines. *Behaviour,* 10, 255–308.

ARCHER, J. (1976) The organisation of aggression and fear in vertebrates. *Perspectives in Ethology Vol. II* New York; Plenum. 231–298.

BROUGH, T. (1968) Recent developments in bird scaring at airfields. In: Murton, R. K.; Wright, E. N. (*Eds*) The Problems of Birds as Pests. London; Academic Press. 29–38.

COLES, R. B.; LEWIS, D. B.; HILL, K. G.; HUTCHINGS, M. E.; GOWER, D. M. (In press). Directional hearing in the Japanese quail (*Coturnix coturnix japonica*). II. Cochlear physiology. *Journal of Experimental Biology.*

FENTRESS, J. C. (1968) Interrupted ongoing behaviour in two species of vole (*Microtus agrestis* and *Clethrionomys britannicus*). *Animal Behaviour*, 16, 135–167.

FRINGS, H.; FRINGS, M. (1967) Behavioural manipulation (visual, mechanical and acoustical). In: Kilyore, W. W.; Douth, R. L. (*Eds*) Pest Control: Biological, Physical and Selected Chemical Methods. Academic Press, New York. 387–454.

GIBAN, J. (*Ed.*) (1962) Colloque sur les Moyens de Protection Contre les Espèces d'Oiseaux Commettant des Dégâtes en Agriculture. *Annales Epiphyties,* 13.

HINDE, R. A. (1954) Factors governing the changes in strength of a partially inborn response, as shown by the mobbing behaviour of the chaffinch (*Fringilla coelebs*). I: The nature of the response, and an examination of its course; II: The waning of the response. *Proceedings of the Royal Society of London B,* 142, 306–358.

HINDE, R. A. (1960) Factors governing the changes in strength of a partially inborn response, as shown by the mobbing behaviour of the chaffinch (*Fringilla coelebs*). III: The interaction of short-term and long-term incremental and decremental effects. *Proceedings of the Royal Society of London B,* 153, 398–420.

HINDE, R. A. (1970) Behavioural habituation. *Short-term Changes in Neural Activity and Behaviour.* University Press, Cambridge. 3–40.

KLING, J. W.; STEVENSON, J. G. (1970) Habituation and extinction. *Short-term Changes in Neural Activity and Behaviour.* University Press, Cambridge. 41–61.

LEGER, D. W.; OWINGS, D. H. (1978) Responses to alarm calls by Californian ground squirrels. Effects of call structure and maternal status. *Behavioural Ecology and Sociobiology,* 3, 177–186.

MARLER, P. (1955) Characteristics of some animal calls. *Nature,* 176, 6–9.

MARLER, P. (1956) The voice of the chaffinch and its function as a language. *Ibis,* 98, 231–261.

MILLS, A. W. (1972) Auditory localization. *Foundations of Modern Auditory Theory.* New York; Academic Press. 301–348.

MOORE, B. C. J. (1977) Introduction to the Psychology of Hearing. London; Macmillan.

MORGAN, P. A.; HOWSE, P. E. (1973) Avoidance conditioning of jackdaws (*Corvus monedula*) to distress calls. *Animal Behaviour,* 21, 481–491.

MORGAN, P. A.; HOWSE, P. E. (1974) Conditioning of jackdaws (*Corvus monedula*) to normal and modified distress calls. *Animal Behaviour,* 22, 688–694.

MORTON, E. S. (1975) Ecological sources of selection on avian sounds. *American Naturalist,* 109, 17–34.

RYDÉN, O. (1978a) Differential responsiveness of great tit nestlings, *Parus major,* to natural auditory stimuli *Zeitschrift fur Tierpsychologie,* 47, 236–253.

RYDÉN, O. (1978b) The significance of antecedent auditory experiences on later reactions to the seeet alarm-call in great tit nestlings, *Parus major. Zeitschrift fur Tierpsychologie,* 47, 396–409.

SHALTER, M. . (1978) Localization of passerine seeet and mobbing calls by goshawks and pygmy owls. *Zeitschrift fur Tierpsychologie,* 46, 260–267.

THOMPSON, R. F.; SPENCER, W. A. (1966) Habituation. *Psychological Review,* 73, 16–43.

THOMPSON, T. (1969) Conditioned avoidance of the mobbing call by chaffinches. *Animal Behaviour,* 17, 517–522.

WOLDA, H. (1961) Response decrement in the prey-catching activity of *Notonecta glauca. Archives neerlandaises de zoologie,* 14, 61–89.

Prospects for making acoustic super-stimuli

J-C Brémond
Centre National de la Recherche Scientifique
Laboratoire d'Ethologie Experimentale
France

Summary

Many birds are able to modify parameters of their acoustic signals in order to express the intensity of particular motivational states, such modifications are perceived by other individuals. In some cases the birds perceiving the altered signal react more strongly than they would have done to the normal version. Once the relevant parameters of the signal have been identified, then it has been shown that we can artificially manipulate the normal signal to create acoustic super-stimuli for attracting or repelling birds. Such super-signals are not super-complicated to synthesise; indeed often they are simpler then the normal call in this respect. However, there is a wide variation in the parameters which different species use in order to signal the intensity of a given motivational state. Much more research is needed if we are to establish any general pattern. This paper examines some examples of such research.

The study of bird vocalization has mostly concerned songs relative to the information they contain about the species identity of the singer. Such work has shown that although in most cases manipulation of the parameters of the signal does not modify this information, the quality of the subject response can differ greatly from the response to the unmodified song. In such cases we generally observe a diminution of the effect of the signal. The purpose of the paper is to consider whether there are modifications that have the contrary effect i.e., that enhance the response? If this possibility exists, then it is probable that birds use such song types normally in the wild. In this way they could, for example, transmit information about the level of motivation, or "mood", of the emitter and thereby increase the efficiency of the communication. It is possible that these critical features of the song, when they have been detected and selected, could be manipulated by the experimenter to build acoustic super-stimuli for attracting or repelling birds.

In order to discover if such behavioural enhancement exists in natural conditions we have to analyse the same signal given in different contexts. First, let us look at the song of the European robin (*Erithacus rubecula*). One of the functions of this song is to claim the ownership of a territory and threaten a potential intruder. If the intruder ignores the warning and enters the territory then the owner modifies his song to make it more threatening. He obtains this effect by singing at a lower sound level (muted song), enhancing the acoustic energy in the upper frequencies, increasing the number of notes per unit of time and extending the total duration of the song. For the listener the result is a higher pitched, more rapid song, with minimal monotony. The robin is able to grade these modifications and probably inform the listener about his level of aggressiveness. In an analogous situation the song-thrush (*Turdus philomelos*) gives a similar muted song which is also compressed in time. The motivational level of the song-thrush is not expressed in the same way as the robin since there is no continuum. The thrush alternates bouts of normal song and compressed song. The rate of production of the modified song indicates the mood; the greater this rate, the more the aggressiveness. Becker (1977) working with the goldcrest (*Regulus* spp.) has made similar observations. The wren (*Troglodytes troglodytes*) however, behaves differently. It modifies neither the content nor the pitch of the song but we have observed that it diminishes the interval between each song in order to express heightened aggression. On the other hand, aggression is diminished by increased spacing and by using shortened (i.e., incomplete) songs. This is often the case near the boundary of the territory where the bird is in motivational conflict between attack and retreat. Lein (1978) showed that the repertoire of the chestnut-sided warbler (*Dendroica pensylvanica*) contained two types of song one of which seemed to be particularly important in relation to rivalry with other males. Switching between these two types of song does not occur in a random fashion. There is a clear relation between the changes made and the external circumstances in which this switching takes place. This bird may also vary his song by singing incomplete or muted versions or either of these types. During a fight he gives what Lein termed "jumbled" song, the constitution of which is similar to the aggressive song of the robin and the thrush. The eastern phoebe (*Sayornis phoebe*) has two types of song and changes in the proportion of these two types within a song sequence, express the variations of the emitter's mood (Smith 1969). The indigo bunting (*Passerina cyanea*) introduces high pitched sounds into his song during fighting encounters (Thompson 1968).

These few examples concern only the song and they show that some birds are able to transmit information concerning the intensity of particular motivational states. The modern tendency is to accept that this ability probably exists in all species but particularly within those having a well developed social life. There are, however, numerous ways of achieving this and it is not possible to predict the method a particular species will employ; each species uses its own method. Nevertheless in all cases a group of parameters remain without modification to ensure the transmission of information concerning species identification and motivational state whilst another group is altered to express the level of that motivational state. This dichotomy is possible with complex signals such as songs, but what happens with simpler and shorter signals such as calls? Is it still possible? Most calls are used in situations connected with "discomfort", food, contact and/or courtship, alarm and threat. All these calls need to be expressed with graduation in order to achieve the most efficient communication.

Let us look at some examples. Goslings taken from their parents give distress calls and if the environment is a strange one they give a greater number of calls (Lamprecht 1977). Chickens behave in essentially the same manner, thus when chicks are left alone in a dark room their distress calls become longer and twice as frequent as when they were in the light (Guyomarc'h 1966). Conversely in safe, familiar situations they emit a special call, the twitter, whose rate of emission is correlated with the welfare of the emitter (Guyomarc'h 1975). In many situations where the activity of robins and wrens is increased, calls of "excitement" are given. The rhythm of the emission of these calls is proportional to the nervousness of the birds. Hens give a food call that attracts chicks. When the chicks are sleeping in comfort under the hen's feathers then she uses faster and louder food calls in order to persuade them to come out (Guyomarc'h 1975). In most passerine birds the rhythm of food begging calls of the young varies with their hunger and adults are very sensitive to these variations. The house-sparrow (*Passer domesticus*) has four types of contact calls. All are used during courtship. The more intense the courtship the more frequently they are used (Rudrauf 1976). Nidifugous birds use contact calls when the young are looking for their parents and by the parents when they are gathering their young together. Paired birds also use localization calls when they are separated and the rhythm of calling is proportional to the duration of the separation. At nightfall mallards (*Anas platyrhynchos*) exchange acoustic signals in order to gather together before leaving the place where they have spent the day. The call they use is at first spaced out and then becomes more frequent, the rhythm of its emission becoming faster until they have almost finished grouping together and flight is imminent. Analogous behaviour of progressive movement coordination is to be found with greylag geese (*Anser anser*).

The repertoire of the house-sparrow contains four calls that are associated with danger. The choice, association and proportion of the calls expresses the emotion of the bird (Rudrauf 1976). The sounds named "distress calls" of *Laridae* and used by Busnel and Giban (1965) to scare gulls were recorded from two or three birds and broadcast without any modifications. An analysis of the calls of each of these birds shows that the sequences were composed by the birds by alternating distress and alarm calls. Has this natural alternation and proportion a special meaning? Does this explain why some sequences were more effective than others? No experiments to resolve these questions have been done. The Pygmy owl (*Glaucidium passerinum*) has a call that, when emitted in a sequence, has an alarm function. The greater the danger the more they increase the rhythm and duration of the call sequence (König 1968). Orange-chinned parakeets (*Brotogeris jugularis*) emit a harsh sound whose loudness is proportional to the degree of anxiety when they see unusual things or when a conspecific approaches the nest site (Power 1966).

Threat is always expressed with gradation. House-sparrows have two calls for this which are subject to loudness and rhythm modifications to ensure expression of graduation of mood (Rudrauf 1976). During fighting, the coot (*Fulica atra*) uses threat calls with increasing pitch (Koronowski 1957). Kermott and Oring (1975) have observed with sharp-tailed grouse (*Pedioecetes phasianellus*) that the intensity and rapidity of call delivery may correlate with level of aggressive motivation. Other grouse calls vary greatly on a continuum depending upon the situations in which they occur. The vermillion flycatcher (*Pyrocephalus rubinus*) emits a special call when the bird becomes very active or approaches a rival. The emission rate increases with increasing motivation and the meaning (activity/aggressiveness) depends on the context (Smith 1967). As the repeated vocalization of the eastern king-bird (*Tyrannus tyrannus*) gets harsher, attack becomes more probable (Smith 1966).

It would be easy to extend this list of examples of signal variation. This sample serves to show that the ability to signal the level of particular motivational states is to be found in both the songs and the calls of many families. In order to express nuances of meaning each species uses its own individual method of variation. An increase in level of motivation of the bird is accompanied by the preferential use of a type of signal, or changing from one variation to another, or an acceleration of rhythm of emission of certain forms. For the calls the problem is further complicated by the fact that the meaning of a call could be modified by changes in the context in which it is used. These variations are always taken into account by the listener who adapts his behaviour to them. The previous examples are very obvious; there are more subtle possibilities. For instance, Guyomarc'h (1974) has shown that a dominance hierarchy can be based upon such features.

The ability of birds to signal information concerning the intensity of particular motivational states is now well established, but is it possible for the experimenter to manipulate these parameters? In order to modify a signal in a predictable fashion it is necessary that the markers of information within it be known. This condition is not enough; only further experimentation can show if the new combinations are understood by the receiver. Unfortunately, experiments in this field are very few; let us consider some of them.

When we broadcast the territorial song of the European robin within a territory we always elicit a response from the owner; an exploration and attack of the loud-speaker. In some cases the response is not very strong, this may happen when the attack is at the edge of the territory, when the territory is very large, with birds that are less aggressive than others, in certain meteorological conditions and at times of the day when all individuals are least active. If we broadcast this song after having increased the energy in the upper part of the frequency spectrum we always obtain a very strong response from the territory owner. We obtain the same result by increasing the diversity of the song elements. These experiments have demonstrated that it is possible to enhance the effect of the song and that only two parameters have to be altered to achieve this.

The territorial song of the skylark (*Alauda arvensis*) is made with a long sequence of various motifs. Each motif is a trill of an element. The overall pattern is a great diversity of sounds given with regular tempo and some repetition. As a first step Aubin (1978) was able to determine experimentally what were the specific markers in this song. He then built a synthetic song using an electronic generator, which included some of the markers that he had found to be effective in producing aggression in the larks. By repeating exactly the same motif with a uniform and regular tempo and by eliminating the intervals between the elements of the trill, he obtained a song that is simpler than the natural song, that is never given by any lark, but has nevertheless a super-normal effect for eliciting threat and fighting displays from larks in the field.

The common American crow (*Corvus brachryrhynchos*) has an assembly call that has been studied by Richard and Thompson (1978). It is a constant and structured sequence of cawing. Each sequence consists of several bursts of one to ten similar caws, each being uniform within itself in the temporal and numerical properties of the burst. In contrast with this, the crows emit also "unstructured" sequences of cawing. The structure and timing of its elements are variable, the tempo is irregular. The meaning varies greatly but is often in relation to dispersal or mobbing. The experimenters have taken an unstructured sequence which elicits mobbing from crows in the wild and arranged them in sequences like structured cawing. From many different arrangements they were able to recognise that the characteristics which promote assembly appear to be a high and increasing rate of emission. An artificial caw that maximises these two characteristics elicits assembly responses from crows within hearing in a higher (two to four times) proportion of tests than the natural assembly call itself.

Alterations to an acoustic signal must be made very carefully and the effect always submitted to experiment. If super-stimuli can be made then the opposite effect might be produced by the introduction of features that are rejection markers (Brémond 1976).

Another possible way of obtaining the maximum efficiency from a signal is to make it interspecific. Unfortunately this is only a theoretical possibility for all the experiments in which decreasing specificity was artificially introduced, have shown a reduction in the positive rate and quality of the behaviour elicited under these conditions. When the interspecificity is the outcome of natural learning this is, of course, an advantage for the experimenter but it is also something we are unable to modify.

Let us suppose that we know what is the optimal signal for the receiver. The next problem is how to ensure that this signal actually reaches the receiver for there is always degradation during transmission through the atmosphere. It is necessary, therefore, to study such factors as attenuation and reflection of the signal in the context of its use. It is then possible to modify the signal, on the tape, in order to prevent some of these degrading effects. This knowledge can also help us to make a choice between different signals, i.e., to reject those that introduce the risk of poor transmission over a great distance. The positioning of the loudspeaker, the signal level and the quality of the apparatus must all be selected according to what it is useful to transmit.

Conclusion

The results discussed previously suggest there is a good chance that super-normal signals can be created. Super-normal signals are not super-complicated. Such signals emphasise only one or a few, of the characteristics of the normal signal, and like a caricature, they are simpler, easier to synthesise and in some cases easier to reproduce than the natural signal which requires a high fidelity device for broadcasting. In order to achieve this goal, however, more fundamental research is necessary. In all practical applications we must not forget that a signal out of its natural context is nothing; it has no meaning for the receiver, and depending on the context, the effect of the signal can be very variable, as has been shown by the number of crows which gather as a result of the broadcasting of corvid distress calls (Brémond 1973).

References

AUBIN, T. (1978) Etude des parametres physiques responsables du declenchement de comportaments dan le chant de proclamation territoriale de l'alouette des champs (*Alauda arvensis*). University of Besançon; Ph.D. Thesis.

BECKER, P. H. (1977) Verhalten auf Lautäusserungen der Zwillingsart, interspezifische Territorialität und Habitatansprüche von Winter – und Sommer goldhähnchen (*Regulus regulus, R. ingnicapillus*). *Journal für Ornithologie* 118, 233–260.

BRÉMOND, J. C. (1973) La panique on exageration des reactions aversives, en rapport avec la taille du groupe chez les corvides. *Revue Comportement Animal* 7, 93–94.

BRÉMOND, J. C. (1976) Specific recognition in the song of Bonelli's warbler (*Phylloscopus bonelli*). *Behaviour* 58, 99–116.

BUSNEL, G.; GIBAN, J. (1965) Le problème des oiseaux sur les aerodromes Colloque tenu a Nice les 25, 26 et 27 Novembre 1963. Paris; Institut National de la Recherche Agronomique.

GUYOMARC'H, J. C. (1966) Les emissions sonores du poussin domestique, leur place dans le comportement normal. *Zeitschrift für Tierpsychologie* 23, 141–160.

GUYOMARC'H, J. C. (1974) Influence du benzoate d'oestradiol sur la fréquence et la hauteur des cris de contact chez la caille japonaise. *Bulletin Societe Zoologique de France* 99, 743–751.

GUYOMARC'H, J. C. (1975) Les cycles d'activite d'une couvée naturelle de poussins et leur coordination. *Behaviour* 53, 31–75.

KERMOTT, L. H.; ORING, L. W. (1975) Acoustical communication of male sharp-tailed grouse (*Pedioecetes phasianellus*) on a North Dakota dancing ground. *Animal Behaviour* 23, 375–386.

KÖNIG, C. (1968) Lautâusserungen von Rauhfusskauz (*Aegolius funereus*) und Sperlingskauz (*Glaucidum passerium*). *Beihefte Vogelwelt* 1, 115–138.

KORONOWSKI, G. (1957) Beiträge zur Ethologie des Blässhuhns (*Fulica atra*). *Journal für Ornithologie* 98, 318–355.

LAMPRECHT, J. (1977) Environment-dependent attachment behaviour of goslings (*Anser anser*) due to environment specific separation experience. *Zeitschrift für Tierpsychologie* 43, 407–414.

LEIN, H. R. (1978) Song variation in a population of chestnut-sided warblers (*Dendroica pensylvanica*). Its nature and suggested significance. *Canadian Journal of Zoology* 56, 1266–1283.

POWER, D. M. (1966) Agonistic behaviour and vocalisations of orange-chinned parakeets in captivity. *Condor* 68, 562–581.

RICHARD, D. B.; THOMPSON, N. S. (1978) Critical properties of the assembly call of the common American crow. *Behaviour* 64, 184–203.

RUDRAUF, J. M. (1976) Contribution à une étude ethologique des rapports individuels et sociaux existant au sein d'un groupe de moineaux domestiques (*Passer domesticus*) élevés en captivité. Thèse psychosociologie animale. Paris.

SMITH, W. J. (1966) Communication and relationships in the genus Tyrannus. *Nuttall Ornithological Club Publication* 6, 1–250.

SMITH, W. J. (1967) Displays of the vermillion flycatcher (*Pyrocephalus rubicus*). *Condor* 69, 601–605.

SMITH, W. J. (1969) Displays of *Sayornis phoebe* (Aves, Tyramidae). *Behaviour* 33, 283–322.

THOMPSON, W. L. (1968) The song of five species of passerina. *Behaviour* 31, 261–287.

Visual bird scarers: an ethological approach

I. R. Inglis
Ministry of Agriculture, Fisheries and Food
Worplesdon Laboratory

Summary

As the majority of commercial visual scarers rely upon novelty to frighten birds, this paper initially considers the responses of animals to novel stimuli. It is argued that unless reinforced by aversive stimulation such scarers may actually become attractive to birds and the difficulty of providing appropriate reinforcement in the farming context is discussed. It is proposed that a scarer which elicits alarm by mimicking aspects of the encounters pest birds have with predators, should be more effective. Such devices are called bio-visual scarers. By their nature they naturally receive reinforcement from time to time and afford potential for eliciting super-normal alarm responses in the pest species. The paper then reviews the types of bio-visual scarer. A distinction is made between interspecific devices which involve stimuli derived from the predator itself, and intraspecific scarers that use stimuli derived from the alarm behaviour of the pest. Evidence is considered, from the fields of ethology and comparative psychology, which enables suggestions to be made for the improvement of existing bio-visual devices and the development of new ones. Finally a plea for more research is made, particularly into the development of integrated scarers in which the appropriate bio-visual and bio-acoustic stimuli are merged.

Introduction

The aim of this paper is to suggest ways of developing efficient visual bird scarers. I do not propose to discuss such factors as cost or ease of maintenance but rather I will concentrate upon those properties of the device primarily responsible for deterring birds over long periods. Often an anthropomorphic view of the likes and dislikes of birds has contributed to a relatively widespread use of an ineffective scaring technique. For example, despite the fact that such fruit as strawberries and cherries are attacked by avian pests, there has been a belief that the colour red is aversive to birds and this has led to the practice of employing red painted bottles on poles as a deterrent. A far better approach to the problem is, I believe, to examine the mass of data available on all aspects of bird behaviour. In this way intelligent guesses as to what stimuli birds themselves find alarming may be made, and then tested experimentally. The success of bio-acoustic scaring devices has shown how productive such an ethological approach can be. I hope to demonstrate in this paper that an analogous procedure is not only possible but also desirable for the development of visual bird scarers.

As the vast majority of commercial visual scarers rely solely upon novelty to elicit alarm, the first major section of this paper examines the responses of animals to novel stimuli. It is argued that scarers which mimic aspects of a pest species' normal encounters with predators should be more efficient than devices that simply rely upon novelty reactions. The former type of scarer I have called "bio-visual". The other main portion of this paper is a review of the types of bio-visual scarers together with discussion of relevant findings from the fields of ethology and comparative psychology that enable suggestions to be made for the development of new bio-visual devices.

Types of alarming stimuli

The ideas presented here have been derived in the main from experiments conducted in the laboratory and therefore may not necessarily be relevant to birds in the wild. However, I believe that it is valuable to consider the potential limitations of visual scaring devices and that field experiments to assess the real importance of some of these issues should be conducted.

Many of the commercially available visual scarers employ stimuli not normally experienced by birds (e.g., revolving orange vanes) and such "artificial" scarers rely upon the novelty of these stimuli to produce their alarming effects. There is a vast literature concerned with the responses of animals to novel objects (for reviews see Fowler 1965, Berlyne 1966, Eisenberger 1972, Russell 1973, Andrew 1974) which contains a plethora of theories seeking to account for the fact that when faced with such objects animals either approach (i.e., explore) them or withdraw (i.e., flee) from them. Obviously from a bird scaring point of view we wish to enhance any properties of the object eliciting the latter response and diminish properties eliciting the former. However, the factors determining which response is elicited are not all properties of the novel object for novelty requires a background of familiarity which is a function of the prior experience of the perceiving animal. The following brief account employs a cognitive approach to this question. This is just one of the possible theoretical avenues, but one which for a number of reasons (Inglis 1975) I believe to be the most fruitful.

The novel/familiar dichotomy requires that within the animal there must be some sort of comparator, a mechanism by which past and present sensory input can be compared. It has been argued that animals assimilate sensory input into a cognitive map (e.g., Tolman 1948) or model (e.g., Sokolov 1963) of their environment. This assimilated input is then used to modify the animals subsequent behaviour by forming "expectancies" which prime the animal to expect certain sensory inputs in the near or distant future. Thus if previous experience has "taught" the animal that stimulus B usually follows stimulus A, then once stimulus A is perceived it triggers an expectancy of stimulus B. The comparator is the mechanism which compares the incoming sensory input with the expected input. If there is a mismatch (i.e., novelty) then the animal attends to the discrepant stimulus and it is the degree of the discrepancy from the expected that determines the type of response. A little mismatch is thought to produce approach and a large mismatch withdrawal (e.g., McClelland & Clark 1953, Dember & Earl 1957). The initial response to a novel object is therefore likely to be withdrawal but each time the object is encountered, more information concerning it is obtained and the expectancies generated by the modified cognitive map form better approximations to the actual input. Therefore withdrawal will slowly give way to approach and finally to the apparent cessation of response once the object has become part of the familar environment.

Visual scarers whose deterrent effects are based solely upon novelty could therefore become attractive rather than remain aversive. Further, the speed at which this will occur should vary not only with the type of device but also with the degree of prior exposure to such scarers of the birds visiting the area. If a given device is in common use then a farmer erecting one will almost certainly find its repellency of shorter duration than if that scarer

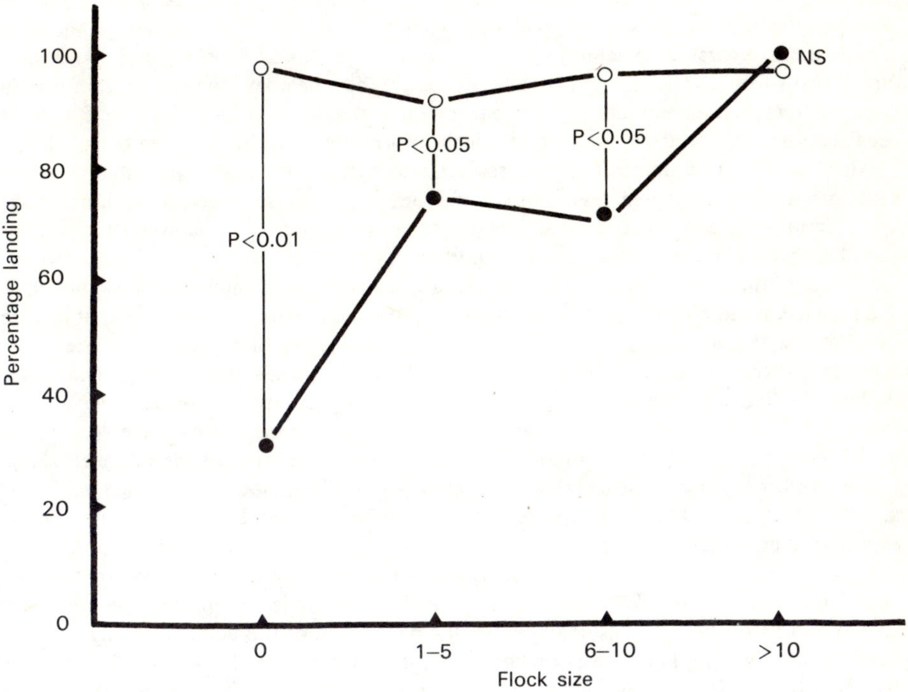

Figure 1. The percentage of woodpigeons circling a feeding area that settled, in relation to the number of woodpigeons already on the area when there was no scarer (○) and when a novel visual scarer was present (●).

had not been used elsewhere in the neighbourhood. It does not require many birds to lose their fear of such devices before all repellency is lost. Fig. 1 gives data which show how the deterrent effect of a novel visual scarer is weakened by the presence of quite small numbers of conspecifics already on the ground.

Once fear of such visual scarers has been lost, it is possible that birds may associate their presence with other, biologically important, features. Farmers will try to protect those crops most vulnerable to bird damage, which are usually the areas preferred by the birds. Thus once a bird has begun to approach the visual scarers it is possible that it may associate their presence with good feeding areas and actively to seek out such devices in the future! How real a problem this is has yet to be assessed, although there is some relevant evidence. Biologists frequently mark the position of study nests with small canes placed some distance from the nest and Picozzi (1975) found that crows (*Corvus corone*) could latch onto this fact, creating a significantly greater level of predation in marked nests than in unmarked nests.

One way to maintain the withdrawal response is to minimise the rate at which information about the novel object can be gained, for example, by exposing the object for short periods only and randomly shifting the position and timing of the exposures. Factors affecting the rate of habituation (i.e., the waning of response to a constant harmless stimulus) are discussed in detail by Slater in this volume and, therefore, I shall not deal further with this point. It is unfortunate, however, that the vast majority of commercial visual scarers, being continuously visible throughout the hours of daylight, appear to have been designed to speed up the rate at which information can be gathered about them.

Another way to maintain a fleeing response to a novel, or indeed any, stimulus is to associate it with subsequent aversive stimulation, i.e., to cause stress. In such cases the first stimulus quickly becomes established as a warning for the subsequent aversive stimulation with the result that the animal learns to avoid the latter by fleeing as soon as the warning stimulus is perceived. There is a vast literature concerning this phenomenon of avoidance conditioning (see Mackintosh 1974 ch. 6) and one fact to emerge is that in such conditioning the warning stimulus may retain its potency for long periods in spite of the obvious fact that once the warning stimulus has been recognised as such the animal never receives the aversive stimulus to reinforce the avoidance response. It has been argued (e.g., Hilgard & Marquis 1940) that the animal forms an expectancy of aversive stimulation contingent upon the warning stimulus and that it is the *omission* of the expected shock which reinforces fleeing. Clearly this expectancy will not be seriously weakened unless the animal "stays around" after perceiving the warning stimulus and then fails to receive the expected aversive stimulation.

In theory the effectiveness of a "novel-type" scarer could be prolonged by converting it into a warning signal. However, there are practical difficulties in creating the necessary spatial and temporal relationships between the perception of the device and any aversive stimulation. Shooting is the most obvious way of providing aversive reinforcement in the farming context for being wounded by shotgun pellets is unpleasant in the extreme and further, as we shall discuss later, the sight of dead or dying conspecifics may also be aversive. Any shooting should not be spatially separated from the scaring devices. Thus a farmer should not place scarers in one area of the farm and then concentrate his shooting effort upon another area for it is possible that the birds will associate the presence of the scarers with the absence of shooting. If so the devices could become attractive "safety signals". Even when shooting is confined within the area covered by the scarers it is doubtful whether the latter will become warning signals for they are continuously visible whether or not shooting is taking place. Thus their sighting provides no accurate information concerning when shooting is to be encountered, i.e., they are very poor predicators of the temporal occurrence of the aversive stimulation and thereby unlikely to become warning stimuli. What is required is a stimulus that appears just before the birds are shot at and disappears afterwards. The possibility of constructing such a device which would also not require any lengthy period of shooting reinforcement by the farmer, is discussed later.

I have argued above that artificial scarers are unlikely to retain their effectiveness for very long and that there might be disadvantages connected with their prolonged use. A more effective approach is, I suggest, the use of natural alarm stimuli. Most avian pests are prey to mammals and other birds at some stage of their life cycles. If we can develop scarers that mimic aspects of encounters with predators then such "bio-visual" devices will to some degree naturally receive aversive reinforcement outside the scaring context on the farm. The birds' response levels to such scarers should remain high throughout a relatively long period, for obviously the individual that does not quickly and persistently respond to a predator is unlikely to pass the genes controlling such behaviour into the next generation. Many species have a genetically in-built tendency to respond to various cues associated with the presence of predators, the strength of these tendencies usually being modified by subsequent learning. Habituation to bio-visual scarers should, therefore, be much slower than to artificial scarers, as has already been reported for bio-acoustic devices (see the paper by Slater in this volume). However, once habituation has occurred the dangers already discussed may still apply.

A further potential advantage for bio-visual devices is that they may be able to produce a response at a greater intensity than that normally shown by the pest. Animals do not usually respond to the entire stimulus pattern that accompanies some biologically important event (e.g., the presence of a mate). Rather, certain aspects are more important than others in triggering the appropriate behavioural response. These important cues are commonly known as "sign stimuli" (Russell 1943) and can frequently elicit the normal behavioural response even when presented in isolation, i.e., without the other stimuli that normally accompany them. For example, a now classic series of experiments (Tinbergen 1953) using models of the heads of herring gulls (*Larus argentatus*) revealed that the characteristics important in eliciting the begging responses of chicks were a long, thin object, pointing downwards, with a red patch near its tip. Further, a red knitting needle with white bands near the tip was found more effective in eliciting begging responses than a model of a normal gull's head. The knitting needle formed what is termed a super-normal stimulus, by combining in an exaggerated form these important visual aspects which are both necessary and sufficient to elicit begging. Presumably gulls do not possess super-effective, knitting-needle bills because eliciting begging from chicks is not the only function of a bill and some of the other functions, e.g., feeding, may not efficiently be served by such a shape, and in any case gull chicks may not *need* to peck more intensely than they do to a normal bill. However, whatever the explanation of super-normal stimuli, the possibility of creating a super-normal alarm stimulus is clearly of immense importance. Such a device might be super-normal not only in the level of the initial response shown by birds perceiving it but also in the time taken before habituation to it becomes complete. The paper by Brémond in this volume discusses work along these lines involving acoustic stimuli and some analogous attempts involving visual stimuli are discussed below.

A review of bio-visual scarers

Two major confounding variables are encountered when trying to assess the efficiency of a bird scarer. First, within any given area there will be variation over time in the numbers of birds attempting to feed. Second, at any given time, areas will differ in their "attractiveness" to birds. It is difficult to devise an experimental procedure that will be relatively immune to fluctuations in these variables. For example, one major component of the "attractiveness" of an area of crop will be the density of available food and this will vary with a number of factors amongst which is the level of bird damage. Therefore, the experimental design used in assessing the efficiency of any scarer must be able to take into account changes in food availability resulting from the operation of the design itself. Preliminary investigations involving computer simulation (Inglis & Huson unpublished work) suggest that the common testing procedure of alternating the positions of control and test area over days may produce particularly misleading results as a consequence of this damage-dependent factor. As there are at present no clear guidelines concerning the validity of different test procedures, it becomes necessary when writing a review of scaring devices to take the published results more or less at face value and simply to try to present a concensus view. This approach is also the only one possible with the numerous reports of an anecdotal nature.

There would appear to be two basic aspects of a natural predator/prey encounter that can be incorporated into a scaring device. First, we can make models of the predator itself and second, we can replicate any warning signals emitted by birds who have sighted a predator. The following account is divided into these two approaches. As the former involves stimuli normally associated with a species other than the pest (i.e., the predator) such devices have been called "interspecific". By contrast the latter approach involves stimuli emitted by members of the pest species and these alternative devices have been termed "intraspecific".

Interspecific devices

Studies of relevance to the design of these scarers may be divided into those involving (a) live or dead raptors or models of such birds, (b) human effigies and (c) simple stimulus configurations that appear to evoke fleeing reactions in a wide range of vertebrate species.

(a) *Avian predators*. Most avian pests are themselves the prey of larger birds and thus it is not surprising that trained raptors have been used as a means of bird scaring. This practice has mainly been confined to airfields (e.g., Van der Heyde 1965; Wright 1965; Heighway 1969; Rodrigues de la Fuente 1971). Blokpoel (1976) concludes from his review of their use that properly trained birds of prey of the appropriate species for the specific location, in the hands of skilled personnel and operated on a daily basis, can be effective in clearing nuisance birds during daylight and good weather. However, as he points out, several falcons are needed in order to have a minimum of one on standby plus a staff of at least two fully-trained personnel. Further, it would appear that the world population of suitable raptors could not sustain the expected required supply if the use of such birds on airfields became widespread.

Bearing in mind the above requirements it is not surprising that trained raptors have not been used in a commercial agricultural context. There is, however, one very interesting experiment by Kenward (1978) who compared the influence of man and goshawks (*Accipiter gentilis*) on woodpigeon (*Columba palumbus*) activity at brassica fields. He found that goshawk attacks, even when repeated and successful, usually failed to keep the pigeons away from the test field for any length of time. Thus, birds resettled immediately after 23% of attacks and returned to feed within the same day after almost 50% of attacks. Indeed, the presence of man was apparently a better deterrent for "human presence almost completely excluded pigeons from the sites". Among several possible reasons for the lack of success with goshawks Kenward suggests that the absence of significant goshawk predation on woodpigeons in England for at least 200 years may have resulted in a lowering of some of the normal anti-predator responses. Further, the recent severe human predation seems to have induced a response to man at least as strong as that to the goshawk.

In order to circumvent many of the problems associated with the use of trained raptors, radio-controlled model aircraft shaped like hawks have been tried on several airfields (e.g., Saul 1967, Blokpoel 1976). Some success has been reported but again skilled operators continuously on call are required. It is not certain whether the realistic hawk shape is in fact necessary since Garrity & Pearce (1973) report success in flushing robins (*Turdus migratorius*) from low bush blueberry fields using an unmodified model aircraft. In this case, however, only the robins were successfully flushed; other species such as waxwings (*Bombycilla cedrorum*) and sparrows (*Ammodramus* sp.) remained, and further, as soon as the 'plane landed the robins returned.

There has been widespread use of models of birds of prey either slung from balloons or suspended on poles. Whilst the results of such work have been variable, in the majority of cases habituation has been very rapid. For example, Meylan & Murbach (1966) tried to protect sunflower seeds from greenfinches (*Carduelis chloris*) by suspending a model of either a hobby (*Falco subbuteo*) or a sparrowhawk (*Accipiter nisus*) beneath a cluster of balloons. After approximately 4 hours habituation was complete. Similarly, A. J. Isaacson and I have assessed the deterrent effect of two commercial scarers, each vaguely resembling a large raptor, within a clover ley. On the morning following their erection on poles 5 m long, a maximum of 125 woodpigeons were observed feeding within 5 m of one of the scarers and birds remained feeding within 50 m of one or other device throughout the entire observation period (0900–1200). Observations over subsequent days confirmed the ineffectiveness of these devices; perhaps not a surprising result in the light of Kenward's (1978) work already discussed.

When trying to account for the ineffectiveness of existing raptor-like scarers it is necessary to consider first, the process of predator recognition and second, the factors determining the type and strength of anti-predator behaviour once the predator has been recognised. The findings of studies concerned with predator recognition are at first sight confusing. Some work suggests that predators are recognised using a few quite simple and general visual cues. Thus, very crude cut-outs of "generalised" raptors can be effective in eliciting fleeing or freezing responses in birds (e.g., Tinbergen 1951, Melzack *et al.* 1959, Melzack, 1961). On the other hand it has also been found that for a given prey, different predators are recognised using species-specific, visual cues which in themselves may be detailed and complex, such as plumage characteristics (Curio 1975). It is probably misleading to consider that the recognition process is based upon either a few, generalisable cues or a series of species-specific, detailed "pictures". Rather it seems likely that these two options are the extremes of a single dimension and that the degree of reliance upon one or other of these approaches will vary with a number of factors.

One important factor may be the age of the prey. Typically, young animals exhibit a lack of specificity in responsiveness, the essential aspects of the situation being quantitative rather than qualitative (Schneirla 1965). As the animal develops, the range of effective stimuli for a given response narrows; this increase in specificity being mediated through some form of learning (see Hinde 1966). Many of the experiments demonstrating the aversive effects of a moving, hawk-like silhouette, have involved young nidifugous birds as subjects. It may be that these young birds were not responding to the hawk silhouette but simply to the sudden change in retinal stimulation which could have been elicited by a number of quite generalised shapes. Several experiments have provided evidence supporting this argument (McNiven 1960, Schleidt 1961). However, not all the data can be explained on this basis. For example, the experiments of Nice & Ter Pelkwyk (1941) and Hartley (1950) used wild, adult passerines as subjects and stationary cut-outs as the test objects.

The degree of prior experience of predators by the prey individual may alter the nature of its recognition process. The more frequently the predator has been sighted, the greater the opportunity the prey has to learn its more detailed aspects. Indeed, Kruuk (1976) has suggested that birds mob predators because this enables them to keep the predator in sight for much longer than would otherwise have been possible without the relative protection afforded by the mobbing group, and thus allows a refinement of the predator recognition process. This proposal highlights the simple fact that detailed observation of a predator requires time. The speed of a predator when hunting could, therefore, also influence the bias of the recognition process. As Hartley (1950 p. 329) has argued "on *a priori* grounds it may be expected that the complex of visual characters by which a fast moving predator is recognised will be simpler than the configuration of characters for a more slowly moving or stationary enemy. There is less time for the necessary avoiding or defensive reaction; and the most effective basis for recognition will be that which depends upon the most obvious characters". Tinbergen (1936) noted that fast moving species like pigeons (*Columba* sp.) and swifts (*Apus apus*) are more frequently mistaken for raptors by other birds than are slower moving species. As Hartley points out, this bias in misclassification is to be expected for obviously "it is better to be hasty than dead". Nevertheless, he emphasises that where possible a more exact recognition process is preferable in order "to prevent the waste of time and energy in needless retreats or futile demonstrations of hostility".

The nature of the visual cues perceived by the prey may control whether many or few cues are necessary for the prey accurately to detect the predator. In other words some cues will be more reliable than others in this respect. The prey will do best if it can seize upon a few simple yet very reliable cues whilst the predator on the other hand should try to hinder speedy recognition by presenting as variable an appearance as possible. This may be one of the possible reasons why many diurnal raptors are dimorphic or polymorphic as opposed to nocturnal birds of prey and carrion eaters which tend to be monomorphic. We might, therefore, expect the prey to recognise the commoner morph more easily and the rarer morph to be the most efficient hunter (see Arnason 1978). It seems, however, that there will be some visual aspects which the predator will not be able to vary, fake or hide if it is to forage efficiently. The body outline and mode of flight whilst hunting (e.g., spėed and wing-beat frequently) would appear to be cues of this type.

It is interesting that some falconers are able to tell at a ġlance whether their raptor is "in the mood" to hunt or not. If the bird is hungry enough to fly at the quarry it is said to be "sharp set". Hamerstrom (1957) compared the tendency of birds to mob a stationary red-tailed hawk (*Buteo jamaicensis*) when it was either well fed or sharp set. Although the bird's behaviour during the test appeared similar whether sharp set or well fed in that it remained on the perch moving its head and shifting position slowly, nevertheless, when sharp set it attracted more mobbing than when well fed. Hamerstrom found it difficult to pinpoint obvious posture differences in the hawk between the two hunger states but suggested that the eyes appeared rounder when well fed and that the head feathers tended to be flatter and the wings held slightly higher when sharp set. This experiment clearly demonstrates the high degree of complexity of which the predator recognition process is capable. It also raises the intriguing question as to why, if the hawk is not mobbed when well fed, it does not fake the satiated signals when hungry in order to facilitate its hunting.

I have suggested that a potential advantage of a bio-visual scarer would be an above normal aversive effect if the device could be made into a super-normal stimulus. A natural question in the present context is, therefore, can we produce a super-normal predator? Unfortunately there appears to be little work relevant to this question.

Super-normal responses have been achieved by exaggerating in some way the sign stimulus appropriate to the natural object. Size is one obvious dimension amenable to such manipulation and indeed Tinbergen (1948) found that the egg retrieval behaviour of the oystercatcher (*Haematopus ostralegus*) was most readily elicited by a model egg for larger than the natural egg. I have found no evidence that the size of a given avian predator has any influence upon anti-predator behaviour. However, Rand (1941) reported that thrashers (*Toxostoma curvirostre*) had a more intense reaction to large snakes than to small ones. Curio (1975) tried varying the degree of contrast, colour and position of the eye stripe on model red-backed shrikes (*Lanius collurio*), this being the important recognition cue, but failed to find a combination that produced a significant increase in the pied flycatcher's (*Ficedula hypoleuca*) mobbing intensity over that given to the normal shrike model.

It has been shown that when a response is influenced by a number of sign stimuli then these can supplement each other. Tinbergen (1951) referred to this effect as the "Law of Heterogenous Summation". Although as Hinde (1966, p. 53) has pointed out "the effect of combining a number of sign stimuli is not necessarily simply additive in an arithmetical sense", nevertheless, it is possible that a super-normal aversive effect might be produced by combining in one model the relevant sign stimuli from several predators. There would appear to be only one experiment of relevance to this suggestion. Curio (1975) after finding that the eyes of owls and the eye stripe of red-backed shrikes were important recognition cues to the pied flycatcher, constructed shrike models with owl eyes and owl models with shrike eye stripes. The response levels to these models were lower than to the normal ones. Curio suggests that the production of such mixed models resulted in their being classified as "novel" birds with a resultant lowering of response. Thus, it is probably incorrect to call these cues sign stimuli as they retained little potency away from their normal visual context (i.e., the rest of the donor's body).

It is important to consider whether the lack of positive results is simply a consequence of the dearth of work on this subject or whether there may be inherent properties of the predator recognition system that preclude super-normal effects. Work on "concept" formation in pigeons is relevant to this issue. Herrnstein & Loveland (1964) demonstrated that feral pigeons (*Columba livia* var.) could distinguish between slides containing people and those that did not. Over 1 200 assorted slides were used which were classified by the experimenters solely on the basis of whether they contained at least one human being or did not. The birds learnt to distinguish the stimuli on this basis and thus apparently were capable of forming a "person concept". Further work has shown that pigeons can distinguish between such classes of visual stimuli as pigeons/other birds (Poole & Lander 1971), natural objects/artefacts (Lubow 1974) and tree/not tree (Herrnstein 1979). The birds were also able to classify correctly new examples of the learned concepts.

There are so many possible features in, for example, "tree" that it is very difficult to determine those the birds are using and which, if any, are necessary and sufficient. In an attempt to circumvent this problem Morgan *et al.* (1976) investigated the learning of an entirely artificial concept. They taught pigeons to distinguish between A and 2 when both were presented in a wide range of type-face. The results did not reveal any necessary or sufficient conditions used in the classification. The concepts were apparently being formed on the basis of a number of overlapping similarities or features, no one of which was essential for classification. It seems likely that the other concepts already mentioned were formed on a similar basis. A concept which lacks necessary and sufficient criteria for its definition has been termed polymorphous (Ryle 1951). If predator recognition is similarly based upon polymorphous concepts then it might be argued that super-normal effects cannot occur as there are no crucial features that can be exaggerated. Lea & Harrison (1978) conducted an experiment in which they taught pigeons an artificial polymorphous concept and then attempted to elicit a super-normal response from these birds. They taught their birds a "two-out-of-three concept". For example, in one of their experiments each visual stimulus had three dimensions; it could be red or green, a circle or a triangle and have superimposed upon it either a black or white shape. The positive features were green, circle and black, thus, if the stimulus possessed any two of these the birds could obtain a food reward by pecking a key when the stimulus was being shown. After the subjects had learnt this task Lea and Harrison for the first time presented a stimulus containing all three positive features and looked for signs of a super-normal response (e.g., a lessening of the response latencies). The birds correctly classified the test stimulus but gave no indication of super-normal behaviour. Thus, addition of a feature superfluous for correct classification did not enhance performance. Although it is unlikely that wild birds would have this sort of "threshold rule" for classification of natural objects, nevertheless, if polymorphous concepts are used then it is possible that the probability of the correct response upon perceiving an object might be correlated with the number of similarities the object has to the "prototype" of the concept, the latter perhaps being based upon an "averaging" of previous experience.

It would be valuable to investigate predator recognition using the above technique in order to discover whether polymorphous concepts are indeed involved. If so, what is the level of classification (for example, "aerial predator", or "owls", or "tawny owl") and what is the process of extinction or habituation of response to these diverse stimuli which have been classified under the same polymorphous concept? That is to say does any decline in response follow "exposures" of the concept, regardless of the individual features comprising the stimulus at any given exposure, or do features drop from the concept classification criteria at a rate dependent upon their individual exposure sequences?

So far I have considered the production of a super-normal response through the use of super-normal stimuli. However, as I will discuss later, there is evidence that corpses, either dismembered or placed in unnatural postures, may elicit alarm responses in birds. It is, therefore, a remote possibility that the presentation of a model predator in an unnatural posture might create more alarm than a similar model presented in a realistic posture. Unfortunately the results of an experiment by Curio (1975) do not support this notion. Realistic models of shrikes and owls which had elicited the strongest mobbing reactions from pied flycatchers became totally ineffective when presented upside-down. Smith & Graves (1978), however, report that the mobbing responses of barn swallows (*Hirundo rustica*) were greater to the head of a great horned owl (*Babo virginianus*) than to the complete owl specimen.

The absence of significant predation pressure from a particular predator can produce a population that fails to recognise the predator when faced with it. For example, the red-backed shrike is absent from northern and southern parts of the range of the pied flycatcher, and Curio (1975) compared the shrike-mobbing behaviour of birds from shrike-free areas with that of birds from shrike-threatened areas. The former group, unlike the latter, showed little or no mobbing to the shrike model although both groups mobbed models of the ubiquitous tawny owl (*Strix aluco*). Curio concluded that the loss of the shrike response in the former group was solely a result of the loss of the shrike recognition mechanism. It therefore seems of little use attempting to scare birds using models of predators not normally found within the area; indeed, as Kenward (1978) has shown, even the responses to *live* unfamiliar predators can be muted.

Clearly much more work needs to be conducted before we begin to understand the mechanism of predator recognition. However, if for a moment it is assumed that we have a device which birds have classified as a predator, we can now consider the factors likely to be important in determining its effective life. Many studies using hand-reared birds have demonstrated a genetic component of anti-predator behaviour (e.g., Melzack *et al.* 1959, Melzack 1961, Curio 1975). On this basis it might be supposed that mobbing, for example, could be triggered again and again by the appropriate sign stimuli with little or no response attenuation. Indeed, Lorenz (1939) recorded that the fear responses of his geese (*Anser anser*) to a hawk-shape did not habituate over a long period. Many other workers, however, have found quite rapid habituation of similar anti-predator responses (e.g., Hinde 1954a,b; Melzack *et al.* 1959; Melzack 1961, Martin & Melvin 1964, Shalter 1978). Habituation rate is dependent upon various procedural parameters (e.g., inter-stimulus interval); however, as Slater discusses these elsewhere in this volume, I propose to concentrate here upon factors more specific to the predator/prey interaction.

Birds do not always flee from raptors. Individuals of many species will approach a predator and mob it whilst emitting calls that induce other birds to join this activity (for a recent discussion concerning the possible functional explanations of such behaviour see Curio 1978). If a bird flees from a predator-type scarer it obviously has little opportunity to gain information about the scarer. Mobbing behaviour on the other hand will provide the individual with such information so that it might be expected that the effective life of the device will be shorter for a species which mobs than for one that flees. However, the act of mobbing may actually enhance fear of the scarer. One hypothesis seeking to explain mobbing is the "Cultural Transmission Hypothesis" (Curio 1978) which in its most extreme form states that a bird learns to fear an object once it witnesses conspecifics mobbing it. There is some experimental evidence for this hypothesis. Curio *et al.* (1978) have shown that "observer" blackbirds (*Turdus merula*) mobbed a non-raptorial bird more strongly as a consequence of witnessing another blackbird mob strongly at the site of presentation, (this bird was in fact mobbing a model of a predator hidden from the observers, the latter seeing only the non-raptorial bird). The strength of the enhanced response was comparable to that elicited by a genuine predator. Further, Curio *et al.* (1978), using the same technique, tutored blackbirds to mob a novel stimulus, a plastic bottle! Cultural transmission of this sort was effective along a chain of at least six birds. It is possible, therefore, that we might improve the efficiency of predator-type scarers by enhancing the mobbing responses of pest species to such devices. It has been reported (Barash 1976, Kruuk 1976) that mobbing responses to a stuffed predator are greatly enhanced if the predator is shown in conjunction with a body or model of the mobbing species. Such an enhancement of the mobbing response could be accounted for by at least two of the "mobbing hypothesis" (see Curio 1978, "Cultural transmission hypothesis" and "Aiding a distressed relative hypothesis"), but whatever the causal and functional explanations may be, adding a body or model of the pest species to a predator-type scarer might serve to increase its effectiveness. A further refinement would be periodic playbacks of the mobbing call of the species concerned.

Exposure to a live predator can revive and/or enhance response levels to model predators (e.g., Martin & Melvin 1964, Shalter 1978). At first sight these findings would appear to be at odds with the suggestion that mobbing serves to sharpen the predator recognition process. However, the birds in these studies may not have been mistaking the model for the predator. It could be that the model was still recognised as such but that the response strength associated with it had been heightened by the prior exposure to the live predator. Hinde (1961) has shown that prior exposure to a strong mobbing stimulus (a stuffed owl) can heighten the normal response to a completely different, weak stimulus (a toy dog) 24 hours later, and that the degree of enhancement is proportional to the length of exposure to the initial stimulus. Whatever the mechanism, this suggests again that scarers imitating raptors will be most efficient in areas containing high concentrations of the relevant raptor. Thus, although Kenward (1979) is probably correct in suggesting that the widespread return of the goshawk to Britain is unlikely to produce a significant reduction in crop damage by its predation upon woodpigeons, nevertheless the increase of this raptor might greatly improve the efficiency of goshawk-like scarers.

Shalter (1975, 1978) showed that a very slight alteration in the position of a predator model revived the responsiveness of captive jungle fowl (*Gallus gallus spadiceus*) and wild pied flycatchers after habituation to the model in the previous position. Thus, predator recognition may incorporate an evaluation of the spatial context which, as Shalter (1978) suggests, could, in part, account for the relative lack of habituation to wild predators as opposed to the rapid habituation rates found in some laboratory experiments (e g., Hinde 1954b). Frequent moving of predator scarers should, therefore, enhance their effectiveness.

The mode of support of the device could affect the level of any anti-predator behaviour shown towards it. Curio (1975) evaluated the effects of adding strange objects to his predator models upon the pied flycatcher's mobbing responses. For example, the placing of a shiny red ball within a body's length of the owl model significantly decreased the level of mobbing to the model. As Curio (1975, p. 75) states "alien stimuli interact with the natural stimulus situation" and thereby appear to reduce the effectiveness of the latter. Suspending model predators from large balloons or mounting them upon conspicuous poles could, therefore, be providing "alien stimuli". Less conspicuous modes of support may lead to greater efficiency.

To summarise the above evidence, it would seem that a raptor-like scarer should, where possible, accurately mimic a particular bird of prey common within the range of the pest species. The deception is more likely to be successful if the device is visible for short periods only and moves rapidly during this time. Coupling the model of the raptor with a model of the pest species and/or the alarm and mobbing calls of that species should enhance its effectiveness. Finally, the mode of support should be made as inconspicuous as possible. Clearly, however, much more work needs to be conducted on the mechanism of predator recognition by birds.

(b) *Human effigies*. The scarecrow has been with us for some considerable time in spite of the fact that it is of little use. Yet man is a major predator of several pest species so that human effigies should in theory be reasonably effective in eliciting alarm. Why they do not in most cases is, I believe, simply because they do not mimic accurately enough the true alarming situation. As already discussed, pigeons at least, are capable of forming complex visual concepts and it would be surprising if they could not distinguish between the little old lady coming to feed them and the sportsman with his gun. For some pest birds, a man with a gun rising from cover is perhaps the stimulus eliciting most alarm and thus a scarer that mimics this, should prove far more effective than an effigy of a man standing with outstretched arms. We have constructed a device that works in conjunction with existing propane bangers to simulate the shooting situation. Field trials have only recently begun, however the preliminary results are encouraging and it is hoped that with the appropriate shooting reinforcement from time to time this scarer will remain effective over a long period.

There is one commercial scarer in human shape that appears to be the exception to the above argument. It is a large, three-dimensional model of a man which periodically shakes its head and slowly waves outstretched arms up and down. In our field trials we have found it the most effective of the commercial scarers so far tested. As discussed in the following section however, the human form could be irrelevant to this device as it may be the nature of the arm movements alone that elicit flight responses through the "Markgren effect" (Markgren 1960).

(c) *Simple stimuli that evoke fleeing*. There are at least two relatively simple stimuli that elicit flight in a wide variety of birds. These are eyespot patterns and what I shall call the "Markgren effect" after the person who first investigated it in any detail.

An eyespot is a circular pattern which resembles the general appearance of the vertebrate eye. Such patterns are found on the wings of many insects, most notably members of the Lepidoptera, and are associated with behaviour in which the eyespots are displayed whenever the insect is attacked by a bird. Such displays can elicit flight responses in the predator. Blest (1957) demonstrated that yellow buntings (*Emberiza citrinella citrinella*) responded to such butterflies with escape reactions and that although the birds' responses waned rapidly in the majority of cases, nevertheless some individuals were conditioned to avoid the insects altogether. Further work involving just the eyespot patterns showed that upon their sudden exposure, escape reactions were elicited in a variety of small passerines. There are two main hypotheses seeking to account for the intimidating nature of such patterns (Blest 1957, Scaiffe 1976b). The first argues that eyespots mimic the eyes of large raptors; the second suggests that these patterns mimic the eyes of conspecifics and that this is alarming because many avian species have frontal threat displays in which the eyes are prominent.

It would appear from a variety of experiments (e.g., Blest 1957, Coss 1972, Scaife 1976b) that the most alarming configuration is one containing two circular eyespots arranged horizontally each containing concentric rings of bright colours. Eyespots painted so as to give a three-dimensional appearance may be better than those consisting of "flat" colours and, in general, large eyespots are more effective than small ones. The pair of eyespots should also appear to stare directly at the bird. Scaife (1976a) has further reported that the context of the eyespots is important. Thus, although the model of a kiwi (*Apteryx* sp.) with inconspicuous eyes was less alarming to chicks than a pair of large eyes in isolation, the model of the kiwi plus large eyes was the most effective. In view of these findings it is surprising that no work has apparently been conducted to develop eyespots as scaring devices. Clearly more data are required about some aspects of their effectiveness, particularly the importance of the context in which the patterns are presented. Nevertheless, the use of such patterns to protect relatively small areas is feasible and A. J. Isaacson and I are currently evaluating the effectiveness of various gadgets incorporating eyespots, in deterring sparrows (*Passer domesticus*) and starlings (*Sturnus vulgaris*).

One of the best ways for a person to scare a flock of birds is to move leisurely towards it whilst slowly moving outstretched arms up and down. Markgren (1960) examined the reactions of various birds to diverse human arm and body movements. He would approach a flock until its members had clearly noticed him (i.e., to the "critical distance") and then retire slightly or move around the flock at this distance, whilst performing a series of movements e.g., jumping, throwing motions, moving on all fours. The slow arm movements already described proved the most effective in eliciting flight and Markgren thought that alarm was created because such movements resembled the approach of a slow flapping predator like an eagle. As already mentioned, a recent commercial scarer incorporates such movements and if Markgren's "eagle-hypothesis" is correct then it is possible that the other attributes of this device e.g., human form, shaking head, may not be essential for its efficiency. Further work on the Markgren effect, particularly variations in rate of "flapping", would, I think, be valuable. Wing beat frequency and general mode of flight, as already discussed, may be cues which are impossible for a raptor to fake whilst still hunting efficiently and thus be among the most reliable for predator recognition.

Intraspecific devices

Birds frequently produce signals that alter the behaviour of conspecifics nearby. Whether such signals serve to inform or manipulate the recipient (Dawkins & Krebs 1978) is not of importance in the present context, as long as by producing devices that mimic these signals we can control bird movements. Such signals can be crudely divided into those that generally elicit approach (e.g., food-finding calls) and those that generally elicit withdrawal (e.g., alarm calls). As already mentioned in the previous section, devices that elicit withdrawal are likely to be the more effective of the two in the long term since the response elicited, by its nature, reduces the rate at which information can be gathered concerning the device, and indeed most research in this field has been concerned with the triggering of withdrawal responses.

Compared with the research concerning the use of intraspecific auditory signals in the control of avian pests (e.g., Busnel & Giban 1960, 1965, Giban 1962, Frings & Frings 1967), there has been very little work investigating the potential of intraspecific visual signals. The experiments that have been conducted may be divided into two broad categories. The first involves the use of corpses or models of the pest species arranged in postures not found in a healthy specimen. The second approach uses models to mimic certain visual displays of the pest species which elicit the desired response (e.g., pre-flight displays).

(a) *The use of unnatural body postures.* The majority of this work has been concerned with the deterring of gulls from airfields. Corpses of gulls have most frequently been used, with varying degrees of success. Saul (1967) reported that bodies of gulls (*Laridae* sp.) displayed in a crucified posture on boards were effective in eliminating lounging flocks of gulls for "as long as the corpse is reasonably tidy, about three months". De Jong & Blokpoel (1966) found, however, that habituation could be very rapid, particularly if the bodies were not frequently moved or were allowed to become bedraggled. Hardenberg (1965) also states that the corpses must be frequently shifted in their position and be kept in a recognisable condition. There appears to have been little detailed observation of the gulls' immediate reactions to such corpses. Frings *et al.* (1955) did note that herring gulls frequently emitted an alarm call at the sight of a hanging dead gull. A recording of this call proved as repellent as the dead bird itself.

The corpses in these experiments were obviously arranged in unnatural postures. However, it seems possible that a deterrent effect might be obtained by using a realistic posture in an unnatural orientation to the prevailing wind, for it is a widespread "rule" within the shooting fraternity that decoys must be placed head to wind if they are not to repel rather than to attract the quarry. Wright (1965) mentions trials in which stuffed gulls in the sitting or standing postures were arranged along runways in various orientations to the prevailing wind. No difference in gull behaviour was observed when the decoys were tail to or cross to wind as opposed to head to wind.

It has been suggested that models of the pest species are far less effective than corpses (Frings & Frings 1967). However, Stout (1975) and Stout *et al.* (1975) reported that model gulls deterred birds from resting areas for as long as the models were employed (up to 8 days). The importance of the unnatural body posture is shown by their finding that whilst a model in the normal upright posture was found to be ineffective, it became very effective when laid on its side. Other effective models were those resembling gulls that had died and dropped to the ground. It was necessary to make the models almost perfect three-dimensional replicas. Saul (1967) on the other hand obtained good results using two-dimensional facsimiles of gulls. It is interesting that only the adults responded directly to the facsimiles although the juveniles would depart if the adults left the area. N. Horton (personal communication) used simple effigies of gulls in the outstretched wing posture in an effort to deter gulls from various portions of a refuse tip. He obtained a short-lasting deterrent effect (i.e., 24 hours) when the effigies were used on the gulls' resting areas but no effect when used on feeding areas. This highlights the danger in extrapolating from successful deterrence on resting areas to effectiveness on feeding areas. Indeed, even within the category of resting areas, gulls can show a wide range of site tenacity correlated with apparent site attractiveness. Thus, Stout *et al.* (1975) demonstrated that gulls resting on the seashore were more difficult to shift than gulls resting on runways less than 100 yards away.

Analogous work involving an agricultural pest has been conducted by Murton and his co-workers (Murton 1970, 1974, Murton *et al.* 1974). They investigated the effects of woodpigeon decoys upon the behaviour of that species and in particular the circumstances under which decoys would result in the greatest shooting success. In the course of this work it appeared that certain decoys might provide a deterrent to other pigeons. Various numbers of bodies with either open or closed wings were laid out and birds passing overhead were watched for a positive response (i.e., dipping, circling, attempting to land, or actually settling). Whilst it was found that the initial positive response rate to open-winged bodies was equivalent to those for bodies with closed-wings, nevertheless with closed-winged decoys 54% of responding birds actually settled whilst only 4% of responding birds would settle with the open-winged decoys (Murton 1974). It was also noted that wings alone would produce the same order of effect as bodies with open wings.

Following on from this work A. J. Isaacson (personal communication) evaluated the effects of woodpigeon wings on damage levels within a clover ley. Forty-nine single wings were placed five yards apart in a grid pattern and left for 78 days. At various times throughout this period damage estimates were made within the experimental area and a control area of the same size. Fig. 2 presents the results. It can be seen that less damage occurred on the experimental area (NB. the apparent decline in damage after day 50 was caused by the new clover growth). Hunter (1974) placed models of woodpigeons in the open wing posture over a two acre cabbage field. At four weeks this field had suffered less damage than two similar adjacent fields although after a further week damage was similar on all three fields.

With this work, as with the gull studies, it is difficult to isolate those aspects of the decoys most instrumental in eliciting the fleeing response. There are several possibilities.

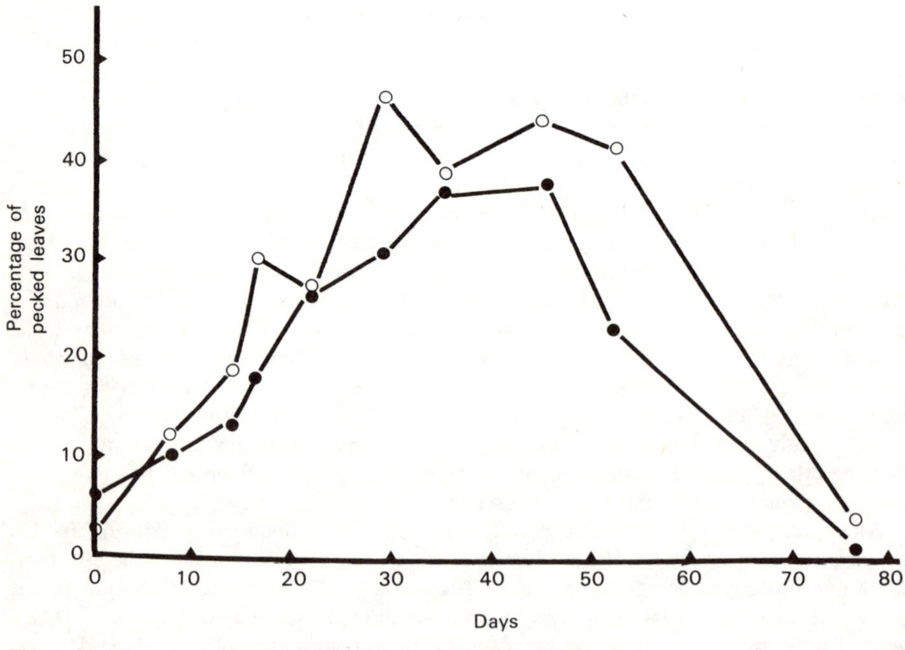

Figure 2. The mean percentage of clover leaves which were pecked within an area covered by a grid of 49 woodpigeon wings (●) and a control area of the same size (○) within the same ley. At no time throughout the experiment did the clover density in the two areas differ significantly.

The unusual open-wing posture may have resulted in the birds not classifying the decoys as bodies but rather as novel objects. This would appear improbable for, as already discussed, feral pigeons at least are apparently capable of forming complex visual concepts and if woodpigeons have the same order of ability then it seems unlikely they would fail to recognise the decoys as woodpigeon bodies. Assuming that the approaching birds do recognise the decoys as conspecifics, then there remain several possible explanations that could account for the flight response. The discrepancy between the abnormal body posture of the decoys and the expected range of postures normally exhibited by woodpigeons on the ground, might *per se* be sufficient to cause alarm. As a result of the unnatural posture the open-winged decoys might be recognised as corpses more easily than the closed-wing decoys. The tendency for predators to indulge in surplus killing (Kruuk 1972, Nunn *et al.* 1976) suggests that the sight of a corpse might be expected to elicit flight as it could be dangerous to remain in its vicinity. On the basis of this argument a fresh corpse should be more alarming than a partially decomposed, although still recognisable, body. Finally, birds may fail to settle near open-winged decoys because these present some intraspecific sign stimuli eliciting flight not shown by closed-winged decoys. The obvious candidates for such sign stimuli are the white wing marks, clearly visible in the open-winged bodies but not seen in the closed-winged decoys. If it is this feature rather than the overall body posture that is important then, as Murton (1974 p. 229) states, "there is scope for experimentation to determine whether the white mark, and its size and distribution relative to the wing could be exploited as a super stimulus." The evidence however is inconclusive and a series of experiments is now being conducted in an attempt to distinguish between the above possibilities. Some prototype devices seeking to mimic the flashing of the wing marks seen as a woodpigeon takes flight have been built and tested, and the results of these field trials will be discussed in the following section.

(b) *The use of normal intraspecific signals*. A type of repellency effect has been created through the use of models of the species concerned even though the models employed did not necessarily mimic any intraspecific alarm signal. The experiments concerned were designed to investigate those features of a feeding flock that make it attractive to conspecifics flying overhead. Krebs (1974) found that great blue herons (*Ardea herodias*) preferred to land with models of herons in the upright hunting posture rather than with hunched roosting models. Also Drent and Swierstra (1977), using models of geese either in the head down posture or in the head up posture, showed that skeins of barnacle geese (*Branta leucopsis*) preferred to land with the model flock containing the greatest number of head down or "grazing" decoys. In both cases the authors suggest that birds flying overhead may estimate the attractiveness of an area for food by the proportion of flock members actually feeding, i.e., hunting or grazing. Therefore, here the attractiveness of an area was lowered not through the use of alarm signals but through the absence of signals predictive of high food density.

Two other studies have involved deliberate attempts to deter bird pests from landing in areas where they would normally have landed, through the use of artifacts simulating possible visual alarm signals. Dark-bellied brent geese (*Branta bernicla bernicla*) sometimes feed on grass and cereals and can cause severe damage. Once a flock has started using a particular field it tends to return each day until the sward is too short for efficient grazing. By using models of brent in various postures Inglis & Isaacson (1978) sought those postures that would deter geese from landing in their usual fields. The most effective model was that simulating the posture shown by the geese when alarmed; the body is angled upward, the neck extended vertically and the beak pointed slighly upward. Just before flighting the geese begin head-shaking, a rapid slight side-to-side vibration of the head, keeping the neck stiff. Raveling (1969) has studied similar pre-flight movements in Canada geese (*Branta canadensis*). The models of geese in this pre-flight/alarm posture were made so that the heads would vibrate slightly when blown by the wind. Two flocks of models were always placed in the test field so that, as well as obtaining a preference measure from the numbers of skeins deterred from landing, it was also possible to judge preferences by noting which "flock" the landing geese joined. When faced with a field containing a flock of alarmed models in addition to a flock of grazing models, only 17% of skeins landed and all joined the grazing flock; when the field contained a head down flock and a head up (i.e., walking posture) flock, 92% of skeins landed, these dividing about equally between the two flocks. We also used flocks containing various combinations of the postures but found that the most effective deterrent was a flock containing all pre-flight alarm postures. Unfortunately these models are fragile and time-consuming to make. In an effort to provide a more viable scarer for farmers, silhouettes of brent in the alarm posture have been made. These can be quickly cut from weatherproof hardboard and it is relatively simple to pivot and spring the head. Some pilot trials have been conducted but the data are insufficient at present for a conclusion to be reached on their effectiveness.

When woodpigeons take off from the ground, the vertical oscillations of the white wing marks are immediately apparent, and may be sign stimuli eliciting flight in conspecifics nearby. Attempts have been made by A. J. Isaacson and myself to simulate this apparent flashing of the wing marks using simple wind-driven devices. Various prototypes have been constructed, most consisting of a number of darkly coloured, rotating vanes with white lines painted on the vanes in such a way that, upon rotation, white marks appear to oscillate vertically. Field trials have shown that such devices effectively deter woodpigeons for about three weeks. Fig. 3 gives the result of one such trial. The test site was a 7 acre portion of a 36 acre pasture that appeared to be a preferred feeding area. A prototype scarer was placed in this area and on subsequent days observations were made on the numbers and positions of pigeons in the field. Fig. 3 shows the mean number of birds in the 7 acre area expressed as a percentage of the total number of pigeons present. The data has been broken into periods with (a) no scarer, (b) scarer not rotating and (c) scarer rotating. The importance of rotation in production of the significant deterrent effect is clearly seen; the mere presence of a non-rotating device having no significant effect. After day 30 the birds rapidly habituated to the rotating device.

Although these prototypes show promise, nevertheless when compared with the efficiency of commercial scarers of similar design but lacking the white marks, they have not always proved significantly better in reducing damage. Thus although we have attempted to mimic the flashing wing marks we may have produced a device which to pigeons (a) does not resemble in any way this situation but is simply a "novel" device, (b) does resemble the situation but not accurately enough or (c) does accurately mimic the situation but the wing marks are not sign stimuli for flight. It has been noted that a sudden rotation from rest by a device will produce flight in birds already on the ground. Perhaps, a model which only periodically rotates would be more effective in simulating a flock suddenly taking flight and work is in progress to test this idea. The further experimentation

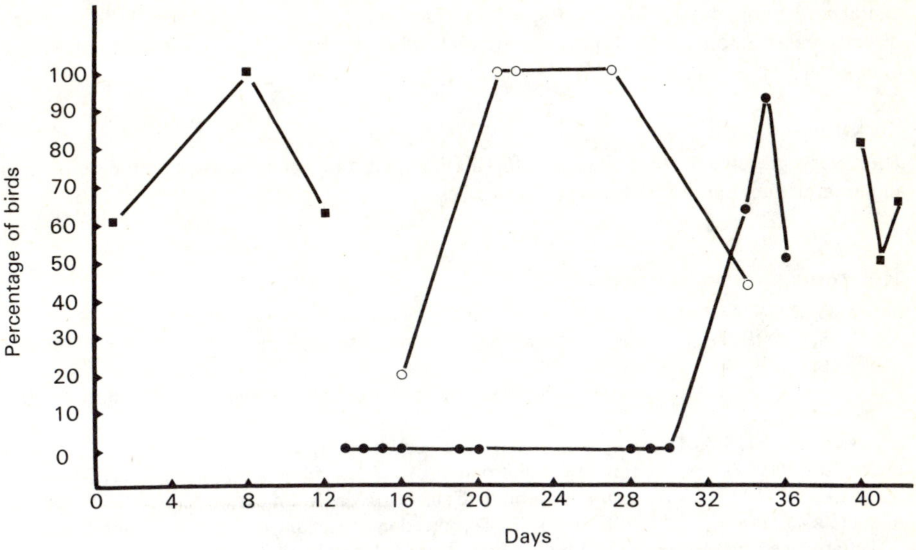

Figure 3. The mean percentage of settled woodpigeons which were within the test area when the wing-bar scarer was absent (■), stationary (○) and rotating (●). See text for further details.

with woodpigeon corpses should shed more light upon the importance of the wing marks; it may be that the sudden expansion of the white neck band seen when woodpigeons are alarmed (Murton & Isaacson 1962), is more important in triggering flight.

Conclusion

In this paper it has been argued that a detailed knowledge of the behaviour of pest birds could lead to the development of bio-visual scarers of greater efficiency than devices that rely upon a novelty factor to produce alarm. This greater efficiency will stem in the main from the natural occurence of reinforcement for the stimuli of the bio-visual scarer. Further, however, such devices offer the potential for triggering super-normal alarm responses. It is possible that super-normal effects may only occur with intraspecific bio-visual scarers. As we have seen, there is no selective advantage for a predator in reducing the ambiguity of the signals it provides for the prey and there may be no critical cues by which a predator is recognised. On the other hand, with the intraspecific communication of alarm, ambiguity of signal must be minimized whatever the function of such calls may be (see Harvey & Greenwood 1978). Such signals, therefore, seem more likely to involve relatively simple sign stimuli that can be manipulated for super-normal effects.

Although in this paper I have concentrated upon visual scarers this does not mean that such devices are necessarily separate from bio-acoustic scaring techniques. The next step should be the development of integrated scarers where the appropriate bio-visual and bio-acoustic stimuli are used together. I have hinted at only a few of the possible instances where such an approach could be employed. Unfortunately, in spite of the potential of bio-visual and bio-acoustic devices, there appears at present to be relatively little research directly based upon an ethological approach to bird scaring. I hope that this situation will change.

Acknowledgements

I am very grateful to John Lazarus for his detailed and constructive comments on this paper and Sue Land for drawing the figures.

References

ANDREW, R. (1974) Arousal and the causation of behaviour. *Behaviour* LI, 10–163.

ARNASON, E. (1978) Apostatic selection and kleptoparasitism in the parasitic jaegar. *Auk* 95, 377–381.

BARASH, D. P. (1976) Mobbing effects by crows: the effect of the "crow in distress" model. *Condor* 78, 120.

BERLYNE, D. E. (1966) Curiosity and exploration. *Science* 153, 25–33.

BLEST, A. D. (1957) The function of eyespot patterns in the Lepidoptera. *Behaviour* 11, 209–255.

BLOKPOEL, E. (1976) Bird hazards to aircraft. Canada; Irwin & Co. Ltd. ISBN 0 7720 1086 2.

BUSNEL, R. G.; GIBAN, J. (1960) Colloque sur le protection acoustique des cultures et autres moyens d'effarouchement des oiseaux, Jouy-en-Josas 1958. Paris; Institut National de la Recherche Agronomique.

BUSNEL, R. G.; GIBAN, J. (1965) Le problème des oiseaux sur les aérodromes. Colloque tenu a Nice les 25, 26 et 27 Novembre 1963. Paris; Institut National de la Recherche Agronomique.

COSS, R. G. (1972) Eye-like scemata: their effect on behaviour. University of Reading, Ph.D. Thesis.

CURIO, E. (1975) The functional organization of anti-predator behaviour in the pied flycatcher. A study of avian visual perception. *Animal Behaviour* 23, 1–115.

CURIO, E. (1978) The adaptive significance of avian mobbing. I. Teleonomic Hypotheses and Predictions. *Zeitzchrift für Tierpsychologie* 48, 175–183.

CURIO, E.; ERNST, U.; VIETH, W. (1978) The adaptive significance of avian mobbing. II Cultural transmission of enemy recognition in blackbirds: effectiveness and some constraints. *Zeitschrift für Tierpsychologie* 48, 184–202.

DAWKINS, R.; KREBS, J. R. (1978) Animal signals; information or manipulation. In Krebs, J. R.; Davies, N. B. (*Eds*) Behavioural Ecology. Oxford; Blackwell Scientific Publications, p. 282–309. ISBN 0 632 00285 9.

DEMBER, W. N.; EARL, R. (1957) Analysis of exploratory, manipulatory and curiosity behaviours. *Psychological Review* 64, 91–96.

DRENT, R.; SWIERSTRA, P. (1977) Goose flocks and food finding; field experiments with barnacle geese in winter. *Wildfowl* 28, 15–20.

EISENBERGER, R. (1972) Explanation of rewards that do not reduce tissue needs. *Psychological Bulletin* 77, 319–339.

FOWLER, H. (1965) Curiosity and exploratory behaviour. New York; Macmillan.

FRINGS, H.; FRINGS, M. (1967) Behavioural manipulation (visual, mechanical and accoustical). In Kilgore, W. W.; Doutt, R. L. (*Eds*) Pest control: biological physical and selected chemical methods. New York; Academic Press. p. 388–454.

FRINGS, H.; FRINGS, M.; COX, B.; PEISSNER, L. (1955) Recorded calls of herring gulls (*Larus argentatus*) as repellents and attractants. *Science* 121, 340.

GARRITY, M.; PEARCE, P. A. (1973) A study of robin control in blueberry fields by use of a radio-controlled model aircraft. Report to the Canadian Wildlife Service, Ottawa.

GIBAN, J. (1962) Colloque sur les moyens de protection contre les especes d'oiseaux dommettant des dégéts en agriculture. Versailles. 1961. *Annales Epiphyties* 13.

HAMERSTROM, F. (1957) The influence of a hawk's appetite on mobbing. *Condor* 99, 192–194.

HARDENBERG, J. D. F. (1965) Clearance of birds on airfields. In Busnel, R. G.; Giban, J. (*Eds*) Le probléme des oiseaux sur les aérodromes. Colloque tenu a Nice les 25, 26 et 27 Novembre 1963, Paris; Institut National de la Recherche Agronomique. p. 121–127.

HARTLEY, P. M. T. (1950) An experimental analysis of interspecific recognition. *Symposia of the Society for Experimental Biology, Cambridge* IV, 313–336.

HARVEY, P. H.; GREENWOOD, P. J. (1978) Anti-predator strategies: some evolutionary problems. In Krebs, J. R.; Davies, N. B. (*Eds*) Behavioural Ecology. Oxford; Blackwell Scientific Publications. p. 129–151. ISBN 0 632 00285 9.

HEIGHWAY, D. G. (1969) Falconry in the Royal Navy. In Proceedings of the World Conference on Bird Hazards to Aircraft, Kingston, Ontario, 2–5 Sept 1969. Ottawa; National Research Council of Canada.

HERRNSTEIN, R. J. (1979) Acquisition, generalization and discrimination reversal of a natural concept. *Journal of Experimental Psychology: Animal Behaviour Processes* 5, 116–130.

HERRNSTEIN, R. J.; LOVELAND, D. H. (1964) Complex visual concepts in the pigeon. *Science* 146, 549–551.

HEYDE, J. M. M. VAN DER. (1965) Moyens utilisés pour éloigner les mouettes et autres oiseaux le l'aèroport de Schipol. In Busnel, R. G.; Giban, J. (*Eds*) Le probléme des oiseaux sur les aèrodromes. Paris; Institut National de la Recherche Agronomique. p. 127–130.

HILGARD, E. R.; MARQUIS, D. G. (1940) Conditioning and learning. New York; Appleton-Century-Crofts.

HINDE, R. A. (1954a) Factors governing the changes in strength of a partially inborn response, as shown by the mobbing behaviour of the chaffinch (*Fringilla coelebs*); I. The nature of the response and an examination of its course. *Proceedings of the Royal Society of London B*. 142, 306–331.

HINDE, R. A. (1954b) Factors governing the changes in strength of a partially inborn response, as shown by the mobbing behaviour of the chaffinch (*Fringilla coelebs*): II. The waning of the response. *Proceedings of the Royal Society of London B*. 142, 331–358.

HINDE, R. A. (1961) Factors governing the changes in strength of a partially inborn response, as shown by the mobbing behaviour of the chaffinch (*Fringilla coelebs*): III. The interaction of short-term and long-term incremental and decremental effects. *Proceedings of the Royal Society of London B*. 153, 398–420.

HINDE, R. A. (1966) Animal Behaviour. A synthesis of ethology and comparative psychology. New York; McGraw-Hill.

HUNTER, R. A. (1974) Preliminary practical assessments of some bird scaring methods against woodpigeons. *Annals of Applied Biology* 76, 351–353.

INGLIS, I. R. (1975) Exploratory behaviour of the hooded rat. University of Bristol; Ph.D. Thesis.

INGLIS, I. R.; ISAACSON, A. J. (1978) The responses of dark-bellied brent geese to models of geese in various postures. *Animal Behaviour* 26, 953–958.

JONG DE, A. B.; BLOKPOEL, H. (1966) Hard facts about soft feathers. *Shell Aviation News* 42, 2–7.

KENWARD, R. E. (1978) The influence of human and goshawk (*Accipiter gentilis*) activity on woodpigeons (*Columba palumbus*) at brassica feeding sites. *Annals of Applied Biology* 89, 277–286.

KENWARD, R. E. (1979) Winter predation by goshawks in lowland Britain. *British Birds* 72, 64–73.

KREBS, J. R. (1974) Colonial nesting and social feeding as strategies for exploiting food resources in the great blue heron (*Ardea herodias*). *Behaviour* LI, 99–134.

KRUUK, H. (1972) Surplus killing by carnivores. *Journal of Zoology* 166, 233–244.

KRUUK, H. (1976) The biological function of gulls' attraction towards predators. *Animal Behaviour* 24, 146–154.

LEA, S. E. G.; HARRISON, S. N. (1978) Discrimination of polymorphous stimulus sets by pigeons. *Quarterly Journal of Experimental Psychology* 30, 521–537.

LUBOW, R. E. (1974) High-order concept formation in the pigeon. *Journal of the Experimental Analysis of Behaviour* 21, 475–483.

LORENZ, K. (1939) Vergleichende Ver haltensforschung. *Zoologischer Anzeiger, Leipzig* 12, 69–102,

MACKINTOSH, N. J. (1974) The psychology of animal learning. London, New York, San Francisco; Academic Press ISBN 0 12 464650 6.

MARKGREN, M. (1960) Fugitive reactions in avian behaviour. *Acta Vertebratica* 2, 1–160.

MARTIN, R. C.; MELVIN, K. B. (1964) Fear responses of Bobwhite quail to a model and a live red-tailed hawk. *Psychologische Forchung* 27, 323–336.

MCCLELLAND, D. C.; CLARK, R. A. (1953) The achievement motive. New York; Appleton-Century-Crofts.

MCNIVEN, M. A. (1960) Social-releaser mechanisms in birds – a controlled replication of Tinbergen's study. *Psychological Record* 10, 259–265.

MELZACK, R. (1961) On the survival of mallard ducks after habituation to the hawk-shaped figure. *Behaviour* XVII, 1–15.

MELZACK, R.; PENICK, E.; BECKETT, O. (1959) The problem of the innate fear of the hawk shape. An experimental study with mallard ducks. *Journal of Comparative and Physiological Psychology* 52, 694–698.

MEYLAN, A.; MURBACH, R. (1966) Versuch zum Schutz einer Sonnenblumerpflanzung gegen Grunfinken *Carduelis chloris* mit Raubvogelattrappen. *Ornithologischer Beo Bachter* 63, 74–76.

MORGAN, M. J.; FITCH, M. D.; BOLMAN, J. G.; LEA, S. E. G. (1976) Pigeons learn the concept of an 'A'. *Perception* 5, 57–66.

MURTON, R. K. (1970) Sophisticated "scarecrows" could be the answer. *Arable Farmer,* September, 10–11.

MURTON, R. K. (1974) The use of biological methods in the control of vertebrate pests. In Price-Jones, D.; Soloman, M. E. (*Eds*) Biology in Pest and Disease Control. Oxford; Blackwell Scientific Publications. p. 211–232.

MURTON, R. K.; ISAACSON, A. J. (1962) The functional basis of some behaviour in the woodpigeon *Colomba palumbus*. *Ibis* 104, 503–521.

MURTON, R. K.; WESTWOOD, N. J.; ISAACSON, A. J. (1974) A study of woodpigeon shooting: the exploitation of a natural animal population. *Journal of Applied Ecology* 11, 61–81.

NICE, M. M.; PELKWYK TER J. (1941) Enemy recognition by the song sparrow. *Auk* 58, 195–214.

NUNN, G. L.; KLEM, D.; KIMMEL, T.; MERRIMAN, T. (1976) Surplus killing and catching by American kestrels (*Falco sparverius*). *Animal Behaviour* 24, 759–763.

PICOZZI, N. (1975) Crow predation on marked nests. *Journal of Wildlife Management* 39, 151–155.

POOLE, J.; LANDER, D. G. (1971) The pigeon's concept of a pigeon. *Psychonomic Science* 25, 157–158.

RAND, A. L. (1941) Development and enemy recognition of the curve-billed thrasher, *Toxostoma curvirostre*. *Bulletin of the American Museum of Natural History* 78, 213–42.

RAVELING, D. G. (1969) Preflight and flight behaviour of Canada geese. *Auk* 86, 671–681.

RODRIGUES DE LA FUENTE, F. (1971) Falconry for the control of birds dangerous on airports. Results of three years of practise. Paper for 6th meeting of Bird Strike Committee Europe, Copenhagen, June 1971.

RUSSELL, E. S. (1943) Perceptual and sensory signs in instinctive behaviour. *Proceedings of the Linnean Society of London* 154, 195–216.

RUSSELL, P. A. (1973) Relationships between exploration and fear: a review. *British Journal of Psychology* 64, 417–433.

RYLE, G. (1951) The concept of mind. London; Hutchinson.

SAUL, E. K. (1967) Birds and aircraft; a problem at Auckland's new international airport. *Journal of the Royal Aeronautical Society* 71, 366–376.

SCAIFE, M. (1976a) The response to eye-like shapes by birds. I. The effect of context: a predator and a strange bird. *Animal Behaviour* 24, 195–199.

SCAIFE, M. (1976b) The response to eye-like shapes by birds. II. The importance of staring, pairedness and shape. *Animal Behaviour* 24, 200–206.

SCHLEIDT, W. N. (1961) Reaktionen von Truthuhnern auf fliegende Raubvogel und Versuche zur Analyse ihner AAM's *Zeitschrift für Tierpsychologie* 18, 534–560.

SCHNEIRLA, T. C. (1965) Aspects of stimulation and organisation in approach/withdrawal processes underlying vertebrate behavioural development. In Lehrman, D.; Hinde, R. A.; Shaw, E. (*Eds*) Advances in the study of Behaviour, I. New York; Academic Press. p. 205–257.

SHALTER, M. D. (1975) Lack of spatial generalization in habituation tests of fowl. *Journal of Comparative and Physiological Psychology* 89, 258–262.

SHALTER, M. D. (1978) Effect of spatial context on the mobbing reaction of pied flycatchers to a predator model. *Animal Behaviour* 26, 1219–1222.

SMITH, M. J.; GRAVES, H. B. (1978) Some factors influencing mobbing behaviour in Barn Swallows (*Hirundo rustica*). *Behavioural Biology* 23, p. 355–372.

SOKOLOV, E. N. (1963) Higher nervous functions: the orienting reflex. *Annual Review of Physiology* 25, 545–580.

STOUT, J. F. (1975) Dispersal of birds from the airport environment. *Proceedings of the 10th Bird Strike Committee Europe.* WP8.

STOUT, J. F.; GILLETT, W. H.; HAYWARD, J. L.; AMLANER, C.J. (1975) Dispersal of seagulls in an aerodrome environment. Technical Report 74–324. Kirtland, New Mexico. United States Air Force Laboratory.

TINBERGEN, N. (1936) The function of sexual fighting in birds and the problem of the origin of territory. *Bird-Banding* 7, 1–8.

TINBERGEN, N. (1948) Social releasers and the experimental method required for their study. *Wilson Bulletin* 60, 6–51.

TINBERGEN, N. (1951) The study of instinct. Oxford; Clarendon Press.

TINBERGEN, N. (1953) The herring gull's world. London; Collins.

TOLMAN, E. C. (1948) Cognitive maps in rats and men. *Psychological Review* 55, 189–209.

WRIGHT, E. N. (1965) A review of bird scaring devices used on British airfields. In Busnel, R. G.; Giban, J. (*Eds*) Le probléme des oiseaux sur les aèrodromes. Colloque tenu a Nice les 25, 26 et 27 Novembre 1963. Paris; Institut National de la Recherche Agronomique. p. 113–119.

The role of refuges in wildfowl management

Myrfyn Owen
The Wildfowl Trust
Slimbridge

Summary

The paper reviews the use of farmland by migratory geese in Britain and examines the advantages and consequences of establishing a network of refuges to alleviate the problem of agricultural damage. Each of the five species of native British geese is adapted for feeding on one of the open habitats that existed before man cleared most of the forests for agriculture. Nowadays, three species feed almost exclusively on improved pasture of arable land and the other two increasingly so. Damage is caused mainly to grass and cereals in spring, with occasional incidences of midwinter use of root crops. A refuge system has been established in North America for many years with the alleviation of agricultural damage as one of its aims. Wildfowl Trust Refuges in Britain have also shown that geese can be attracted from private farmland to refuges and that their operation is not unduly costly.

Several probable consequences of refuge creation are discussed, the most important of which are:

(a) the likelihood that refuges will either accelerate the adoption of arable feeding by birds at present on seminatural habitats and

(b) that refuges in some cases may result in increases in overall population levels by decreasing mortality and possibly by increasing the population's breeding potential. These problems can be overcome if they are appreciated at the outset and refuges are properly managed.

Some suggestions are made on the best location, size and shape of refuges as well as possible methods of acquisition and management. The system will not solve all damage problems and possible exceptions are discussed.

Introduction

All species of wild geese in Britain use agricultural land to some extent and most do so in large numbers and so come into conflict with farmers. Swans and some ducks also give rise to complaints but incidences of damage are restricted to a few areas and are not usually serious (Owen & Thomas 1975). This paper examines the problems set by geese on farmland and examines ways in which these could be overcome. The use of refuges, proposed by Owen (1977) is discussed further and some possible problems examined in the light of recent work.

Geese and farmland

Five species of migratory geese visit Britain in numbers and all of these probably did so in prehistoric times, when more than 90% of the country was covered by forest. Geese are birds of open land and each species was adapted to use a particular kind of habitat. Fig. 1 summarises diagrammatically the kinds of habitats used by each species traditionally and the position today after man has cleared most of the forests and claimed the vast majority of the land for agriculture.

Greylag geese (*Anser anser*) have bills adapted for digging and tearing up roots and other underground parts of plants. They were resident on the larger tracts of fenland and inland marshes and probably wintered on coastal areas, feeding on reed and bulrush beds. A migratory population came from Iceland to winter on Scottish estuaries and Irish marshes. Nowadays, the inland marshes have disappeared and agriculture and industry have encroached onto most of our larger estuaries. The species is almost entirely reliant on arable land (including grassland) for its winter feeding. The distribution of greylags has changed slightly, with a retreat from Ireland and increasing concentration in central and eastern parts of Scotland.

Pink-footed geese (*Anser brachyrhynchus*) are well adapted for grazing and they probably occupied the sandy estuaries of the west coast where they grazed fine grasses on the shifting sandflats. They now spend most of the year on arable land and their distribution has changed to the extent that numbers have declined on southern English estuaries and there are large concentrations in central and eastern Scotland.

White-fronted geese (*Anser albifrons*) came from Greenland to western parts of the British Isles where their traditional acid bog habitat was to be found. They fed on underground parts of bog plants and grazed growing leaves in spring. A small number of European whitefronts might have wintered in southern England as they do today. Both species are now on arable land or improved grasslands. The Greenland whitefront is distributed much as it was formerly, it has merely occupied agricultural land close to its traditional haunts.

Barnacle geese (*Branta leucopsis*) have an extremely rapid pecking action which enables them to feed on very short swards. They were found on exposed islands of western

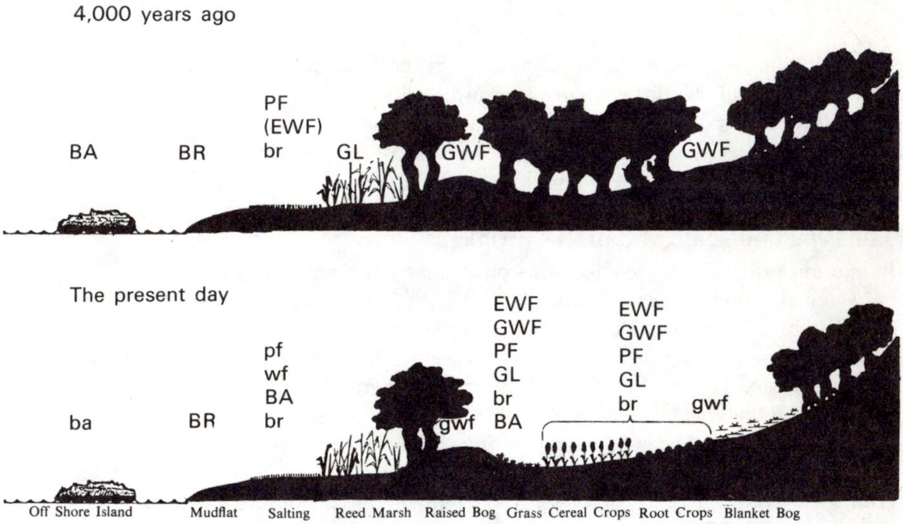

Figure 1. Diagrammatic representation of the British landscape before and after man claimed the major part of the lowlands for agriculture. Capital letters indicate a major habitat, lower case a supplementary one. BA=Barnacle, BR=Brent, EWF=European Whitefront, GWF=Greenland Whitefront, GL=Greylag, PF=Pinkfoot.

Ireland and north-west Scotland, where vegetation was kept short by the effects of wind and sea spray. The geese are still found on these islands but the grazing has largely been improved and is used by farm stock. The numerical distribution has changed radically in recent times, however, with about two-thirds of the Greenland population wintering on the island of Islay in the inner Hebrides, where, formerly, the species was found in much smaller numbers.

Brent geese (*Branta bernicla*) came from Siberia and from Greenland and Canada to winter on the muddy estuaries of the east coast of Great Britain and in Ireland. They fed exclusively on eelgrass (*Żostera*) and algae in autumn and winter, moving to nearby saltings in spring to graze the new growth of grass. The light-bellied race from Greenland (with a few from Spitsbergen) remain on their traditional habitats but dark-bellied brent in southern and eastern England have recently resorted to fields of growing winter wheat and to arable grasslands (St. Joseph, in press).

The bean goose (*Anser fabalis*) used to winter in large numbers, especially in Scotland, but only tiny populations now remain.

Two main points emerge from the above account:

1. The traditional habitats of geese in Britain have largely been taken over for agriculture or forestry. In most cases geese remained on their traditional haunts after they had been reclaimed or moved rather short distances to nearby agricultural land.
2. Although there has been a radical change of habitat for most species their traditional attachments are strong and there have been no consequent major distributional shifts, although the relative numbers in different areas have changed following the change in habits. The present distribution of migratory geese in the British Isles is shown in Fig. 2.

In addition to these arctic breeding species the Canada goose (*Branta canadensis*) is resident in parts of Britain, chiefly central and southern England. The species was introduced as an ornamental bird on large estates but numbers have now increased beyond the capacity of the release points and most of the feeding is on agricultural land throughout the year. (Ogilvie 1977).

The type and scale of damage problems

In autumn greylags and pinkfeet feed on stubble grain and waste potatoes, and the grass on which the whitefront and barnacle feed is of little value to the farmer. There are few complaints at this time. During hard weather in midwinter greylags eat swede turnips which Scottish farmers rely on to feed sheep in winter. Greylags have also been known to eat unharvested carrots. Pinkfeet in Lancashire, where carrots are grown extensively, also began eating the unharvested crop in 1973, having previously taken waste roots left out on fields after harvest. The incident was serious because it took farmers by surprise but there have been only a few isolated occurrences in more recent years. Damage to a high value crop such as carrots can be severe, since a small amount of damage to the roots makes them unmarketable.

It is in late winter and spring that the most serious problems occur, when the geese move to growing cereals and grassland. Brent geese have caused substantial reductions in the yield of winter wheat in Essex in circumstances when their visits coincided with periods of unusually wet weather (Deans, 1979). The combined action of grazing and

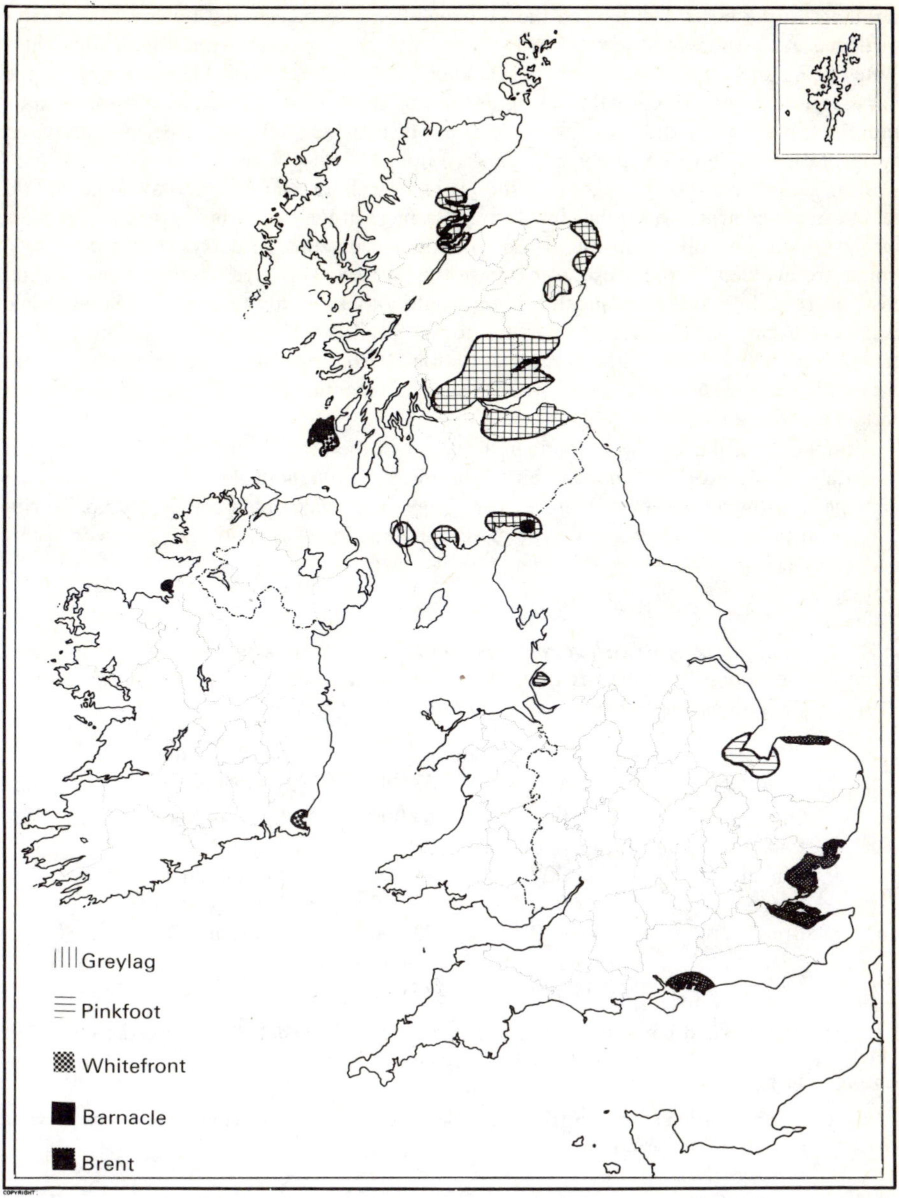

Figure 2. The distribution of geese on agricultural land in the British Isles. The map is not intended to represent a complete distribution of all species, only to indicate areas where damage problems do or are likely to occur. Based on Wildfowl Trust counts.

paddling by the birds' feet caused the growing shoots to be covered by soil, which later hardened and suppressed new growth, leaving the crop open to weed infestation. Pink-footed geese and greylags remain in Scotland until late April or early May and at that time compete with farm stock for the spring growth of grass. Geese select the most nutritious and digestible grasses and concentrate on those fields which have been specially prepared by the farmer to provide 'early bite' for his stock.

Barnacle geese are very gregarious and graze very short swards. On Islay some 20,000 geese are concentrated on rather few farms. Farmers argue that winter grazing depresses spring growth and opens up the sward to allow weeds such as daisy, (*Bellis perennis*), which are avoided by the geese, to colonise the pasture. This gives rise to the additional cost of reseeding more frequently than would otherwise by necessary. Small scale exclosure experiments indicated that the goose grazing reduced April standing crop of grass from 1800 to 2000 kg/ha (Frame & Patton 1976). Their experimental methods were criticised by Owen (1977) but nevertheless substantial loss of grazing for stock undoubtedly occurs.

Complaints of damage were made in a quarter of the areas occupied by Canada geese in Britain (Ogilvie 1977). Since the birds are present throughout the year, they can cause damage to growing and even mature corn. They enter waterside fields and strip the ripe corn from the stalks. They are also accused of competing with stock for pasture in spring and summer.

Table 1

1978–79 populations of British geese and an estimate of those which are on arable land at a time when damage to crops is likely to occur. Data from M. A. Ogilvie, pers. comm., A. K. M. St. Joseph, pers. comm. and Wildfowl Trust counts.

Species	Population Total	Population Arable	Season damage most likely
Greylag	64 000[1]	64 000	March–April
Pinkfoot	66 000[1]	56 000	March–April
Whitefront	25 000[2]	10 000[2]	March–April
Barnacle	39 000[2]	22 000	January–April
Brent	60 000	23 000	February–March
	254 000	175 000	

[1] Greylag and Pinkfoot populations are calculated to have suffered 15% mortality by the spring. This is based on their 19% and 18% respective annual mortalities (Ogilvie & Boyd 1976) and the allocation of losses discussed by Boyd (1956).

[2] Figures for a proportion of the Barnacle and White-fronted Goose populations are estimates based on surveys in previous seasons.

The number of geese involved varies from year to year according to survival and breeding success. The numbers recorded in the current winter (1978–79) are given in Table 1. All but a few of the greylag and pink-footed geese are on arable grasslands or cereals in spring and can be considered to cause damage. Only those barnacle geese wintering on Islay are considered here as a threat to farmers although the increasing

Solway flock may eventually come into conflict (Owen & Norderhaug 1977). Brent geese nest in the high arctic and tend to breed either very well or very badly. In good seasons the extent of use of agricultural areas is much greater than when there are very few young and this is only partly explained by the increase in numbers brought about by good breeding (St. Joseph in press). The numbers feeding on arable fields in 1979 was estimated to be 23 000 (A. K. M. St. Joseph, pers. comm.).

The total number of geese in conflict is small on a national scale but Fig. 2 shows how localised are goose concentrations and problems are serious enough in some areas to require solution. Since it is clear that most of our geese will have to be accommodated on farmland indefinitely, a long-term solution must be found.

Scaring devices give a limited and short-term relief but no scarer has yet been devised which is effective over a long period and with little maintenance. Compensation schemes are costly to administer and liable to be abused. Goose dispersal from favoured areas has been advocated but extremely intensive harassment has as yet proved ineffective in dispersing the concentration of Canada geese on Horicon Marsh, Wisconsin (Gilbert 1977). Roost disturbance may be more effective in some of the smaller British sites but experiments have yet to be carried out.

Because most of the goose species involved have increased in number in recent decades, many agriculturalists advocate substantial culling of present populations. However, in most cases the increases represent recovery after a period of more intensive hunting, a response to a series of mild winters and good breeding conditions.

Refuge creation

The idea of creating refuge areas so that geese can feed without conflict with farmers is not a new one. A system of wildfowl refuges has been established in the United States with the twin aims of securing the future of wildfowl populations and avoiding serious conflicts with farmers. It is now estimated that some 4.5 million hectares are controlled and managed for migrating and wintering wildfowl by federal, state and private agencies in North America (Sanderson 1976). The impetus for this programme came largely from conservationists and hunters but the alleviation of agricultural damage is seen as an important function of refuges. Large numbers of geese have been attracted to such refuges, for example, the Horseshoe Lake Refuge in Illinois attracted 10 000 Canada geese in 1950, soon after its establishment. The number increased to 100 000 in the 1960s (Reeves *et al.* 1968) and 200 000 in the 1970s.

United States refuges are intensively farmed for geese, cereals being the most widely grown crops for winter use and grass and clovers to provide spring grazing. Maize or other cereal crops are in some cases left unharvested to be eaten by geese and some refuges operate a system of share-cropping, whereby the crop yield is divided between the farmer and the birds, in a proportion agreed between refuge managers and the agricultural contractor (Givens *et al.* 1964).

In Britain most wildfowl refuges protect only the roosts and the birds have to forage on nearby farmland. Only a few areas, owned by private organisations, include arable land which is managed in the interests of geese. The two goose refuges established by The Wildfowl Trust succeeded in attracting geese from nearby areas and taking the pressure off local farmers. In both cases the land was kept in agricultural production and management was not radically different from that practised on other farms in the

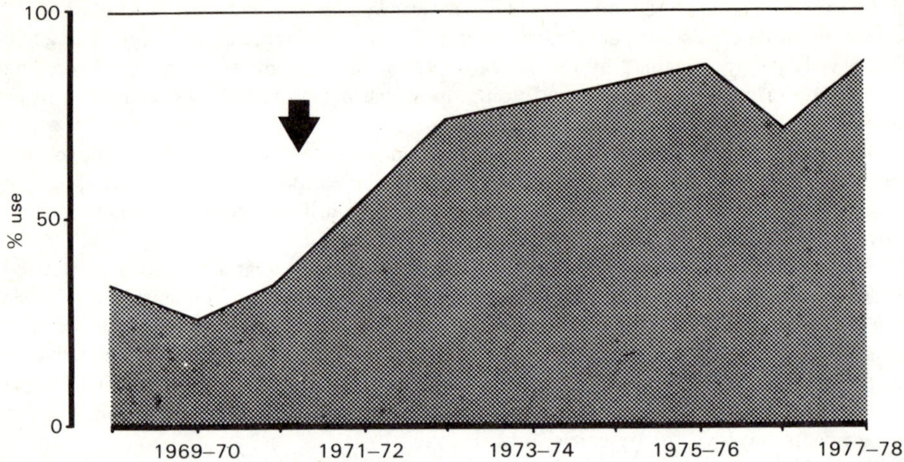

Figure 3. The proportion of time spent by European White-fronted Geese inside and outside the refuge area at the New Grounds, Slimbridge, before and after its establishment (indicated by the vertical arrow). Refuge use is indicated by the shaded area.

neighbourhood (Owen 1977). The proportion of time spent by geese in an inner protected zone managed for geese and adjoining areas over 10 seasons is shown in Fig. 3. The use of refuge land increased from about 35% before establishment to about 80% in later seasons. There was no pressure from adjoining farmers to disturb geese from their land, but goose numbers in the area as a whole declined during the period.

Possible consequences

There is little doubt that properly managed refuges can attract and hold the birds but their establishment could also have unintended consequences which should be borne in mind.

(a) **Disease risk**

Large concentrations of birds increase the likelihood of disease transmission. Outbreaks of diseases such as botulism in North America occasionally causes extreme problems on refuges because of the vast numbers of birds involved (Jensen & Williams 1964). In Britain, however, with our smaller goose populations and less favourable conditions for botulism, the problems are unlikely to be great. The risks are likely to be most severe on roosts, but many roosts in Britain are now subject to protection and already hold very high densities of wildfowl without ill effects.

(b) **Modifications of feeding habits**

Farmland foods are often more nutritious and easily digested than naturally occurring ones and it has been suggested that if crops are provided on refuges, geese will learn to use them eventually to the exclusion of their traditional foods. Thus, if only a proportion of the population were using farmland before refuge establishment, all the others would do so and overspill onto neighbouring private land.

It seems likely that the refuge system in North America has accelerated the adaptation of wildfowl to use farmland for feeding. The acquisition of novel feeding habits is more rapid in areas of great concentrations of birds. For example, the use of farmland by bewick's swans (*Cygnus columbianus*) at the Ouse Washes developed rapidly after a large number of birds began to concentrate on a refuge for roosting.

The danger of creating another farmland goose was the reason for the Canadian Wildlife Service decision not to create inland refuges to accommodate increasing numbers of the greater snow goose (*Anser caerulescens*) on its staging areas in eastern Canada (H. Boyd pers. comm.). Refuges should be used sparingly for species such as the brent goose and carefully monitored so that the birds are not lured inland from the mudflats, where they do no damage.

(c) **Population changes**

One of the chief aims of the United States refuge system was to allow populations to grow so that the harvestable stock increased. This was spectacularly achieved for several Canada goose populations (see Bellrose 1976 for details). Canada goose numbers prior to the creation of refuges had been kept well below the capacity of the habitat by hunting pressure and it was the decline in mortality that accounted for these increases (see Reeves *et al.* 1968). As populations increased in size, more and more refuge land had to be acquired and managed to accommodate them.

While population increases were considered highly desirable in North America, where hunting directly or indirectly finances the refuge programme, in Britain, where populations have already shown considerable increases recently and where there is no great danger of

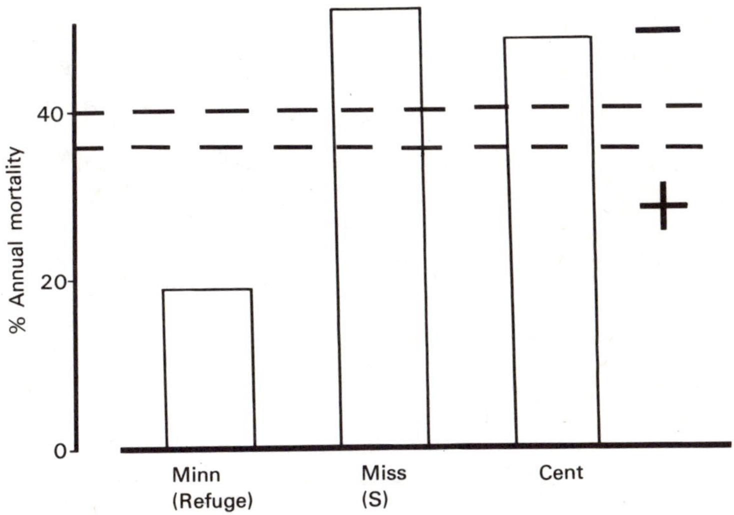

Figure 4. Annual mortality rates of individually marked Giant Canada Geese wintering in Minnesota (mostly on a large refuge at Rochester), farther south on the Mississippi Flyway and in the Central Flyway. Horizontal dashed lines indicate average annual recruitment to populations of large Canada Geese. Based on data in Raveling (1978).

over-shooting, such increases would be considered undesirable by agriculturalists. Are we likely to witness large-scale expansion in numbers if a refuge system were created?

Populations increase when recruitment levels increase or when there is a decrease in mortality, and refuges could have an effect through both of these. Mortality of Canada geese on refuges where shooting was controlled was much lower than that recorded elsewhere (Raveling 1978, Fig. 4) and since a large proportion of geese were on refuges during the shooting season, overall mortality declined and numbers increased.

Most British populations are subject to mortality levels below 20% annually (Ogilvie & Boyd 1976, Ogilvie & St. Joseph 1976, unpublished Wildfowl Trust data) and these are very low by North American standards. Thus any effect is likely to be less spectacular in the British context. Shooting mortality might in any case not be completely additive and protection in some areas might have little effect on overall mortality. It can be seen from Table 1 that damage occurs mainly in spring, outside the shooting season, so that as long as refuge areas were operated normally at other times, mortality would not be affected. It seems unlikely, therefore, that the North American experience following the creation and expansion of refuges would be repeated here.

Evidence on the effect of body condition of geese in spring on breeding success in geese is accumulating. Energy reserves are crucial to breeding in arctic species (MacInnes *et al.* 1974) and these are laid down in March and April when the vegetation is beginning to grow. Fig. 5 shows the condition of female barnacle geese in May and subsequent breeding success. The data are very preliminary but show that on each of five dates females which subsequently bred successfully were in better condition than those which failed (Sign test P=0.031). Barnacle geese wintering on the island of Islay where spring feeding is on improved and in some cases heavily fertilised pasture, produced on average

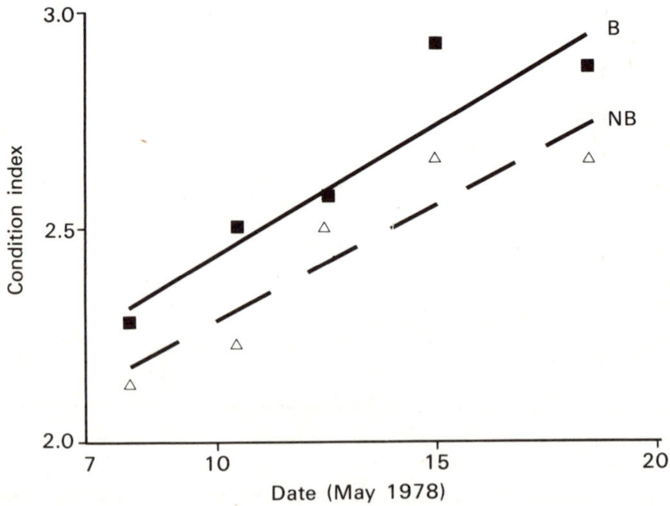

Figure 5. The abdominal profile, an indication of body condition (Owen in press) of female Barnacle Geese in May 1978, which were subsequently (autumn 1978) seen with (B) or without (NB) young. Sign test for direction of difference at each date P=0.031.

15.4±2.0% young in the years 1961–72 whereas those on overgrazed and largely unimproved Inishkea Islands in Ireland averaged only 6.6±1.1%, t test P<0.01 (data from Cabot & West 1973 and M. A. Ogilvie pers. comm.). Boyd (1974) suggested that the difference was due to the better winter and spring feeding on Islay and in view of the data shown in Fig. 5. This explanation seems likely.

The consequence of increase in numbers is predictable; the capacity of refuges will be exceeded and geese will spill over to nearby private land. Fig. 6 shows the use of the Eastpark Refuge, Caerlaverock, by barnacle geese in relation to the use of the Solway as a whole. After the refuge was established, the use of the land by geese more than doubled, but has not increased for the last four seasons, although the time spent by geese on the Solway has gone up by 85%. The creation of the refuge did contribute to the increase, since geese were accommodated there during the shooting season (although the population is protected, most of the mortality is from shooting) and there was a considerable reduction in mortality.

It is probable, then, that the creation of refuges which give a species protection throughout the winter will cause a decrease in mortality and spring refuges an increase in recruitment. The true effect of the latter on species such as greylags and pinkfeet whose relative productivity is declining (Ogilvie & Boyd 1976) is difficult to predict.

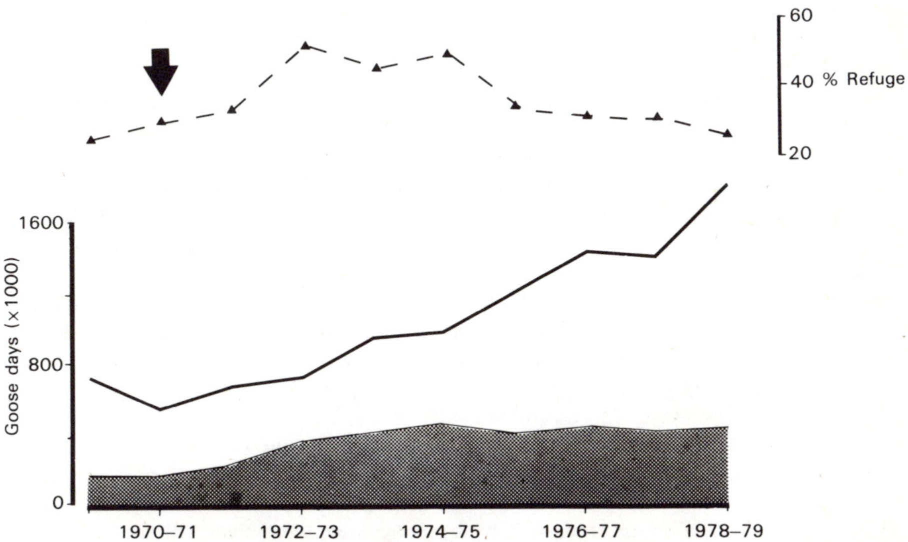

Figure 6. The use of the Eastpark Refuge (shaded) and the whole of the winter range (solid line) by Barnacle Geese over 10 seasons, expressed in goose days ×1 000. The vertical arrow indicates the establishment of the refuge and the dashed line the proportion of the total goose time spent on the refuge.

(d) **Distributional shifts**

In North America the creation of refuges on the flyway has resulted in a phenomenon known as "short-stopping", where geese stop short of traditional wintering grounds because favourable refuge conditions make it possible for them to winter farther north. Increases in refuge concentrations could thus be caused by changes in distribution as well as overall population increases. Raveling (1978) showed that in the giant Canada goose much of the redistribution could be explained by different levels of mortality on different segments of the population rather than by mass changes in wintering areas. Fig. 4 shows that the mortality of the segment wintering in Minnesota, chiefly on one large refuge, is much lower than farther south on the same flyway and in the central flyway.

The refuge population increased because mortality was lower than average recruitment rate and other segments were progressively eliminated. Geese are traditional birds and it is likely that concentration by redistribution is only a major problem in areas where mortality is high and variable from place to place. A number of medium-sized refuges would be preferable to a few large ones in this respect.

Practical considerations

(a) **How much land is required?**

Using the most intensive use of present refuges as a guide, it was calculated that some 7 500 ha of land would be required to accommodate the 140 000 geese which were potentially damaging in 1975 (Owen 1977). Goose numbers have increased since, so necessary land might be between 9 000 and 10 000 ha. This land would not be taken out of agricultural production and we can still conclude that refuge creation is an inexpensive method of tackling the damage problem.

(b) **Size and shape**

The length of perimeter in relation to the area within a refuge is of importance, since influences unfavourable to the birds are usually associated with human activities around the periphery. Thus a circular area is the most effective and the larger the refuge the less will be the edge effect. In practice, few large areas may have other disadvantages (see above) so that units of between 200 and 500 ha are likely to be a reasonable compromise.

(c) **Location**

Because of the traditional attachments of geese, refuges should be sited in areas already occupied by the birds and where damage problems are already severe. Such refuges are likely to be most efficient both in their adoption and usage by geese and in alleviating local problems. A refuge should ideally include a roost and a feeding area and these should be contiguous to allow the birds undisturbed flight lines. The acquisition of land adjoining existing protected roosts would ensure that these conditions would be met.

(d) **Acquisition**

The easiest method of refuge creation is by a management agreement with the landowner. Under such an arrangement, a farmer is paid a sum on an acreage basis in return for surrendering the shooting rights and/or limiting disturbance on fields during the critical period. Leasing land and arranging management through a sub-tenant gives more control over day-to-day refuge activities. The sub-tenant is compensated for the presence of the geese by rent reductions. Ownership and farming of land gives the maximum control over conditions on the ground and since management has to be flexible to allow for seasonal variations in weather and productivity of crops, this is the best possible method. However, it is likely to be the most cóstly, since losses will be made if stock have to be moved or sold according to refuge requirements without regard for agricultural practices or the state of the market. Leasing seems to be a reasonable compromise and is the arrangement by which both Wildfowl Trust refuges are managed.

(e) **Cropping**
The maximum value of a refuge is realised if the whole area is put over to the crop which attracts the birds at the time when damage is most likely. For example, pinkfeet and greylags do no damage when feeding on stubbles or waste potatoes so that refuges for those species should concentrate on providing good quality grass in spring. Because of agricultural considerations, cropping may have to be varied, but where cereals are grown, they should either be winter wheat, which provides useful spring forage, or undersown, so that the grass sward is available for spring use.

(f) **Exceptions**
Refuges are most effective in safeguarding against regular damage which is predictable in its timing, and this might be called "chronic" damage. Occasional "acute" incidences are likely to occur from time to time, such as the localised damage to carrots by pinkfeet or possibly to winter wheat by brent geese. Cooperation between biologists and farmers coupled with intensive campaigns of scaring and disturbance on vulnerable crops have proved effective in preventing the large-scale recurrence of incidences in recent years.

The introduced Canada geese are scattered and cause damage over a long period in spring and summer. Man has encouraged their expansion by translocation and damage could be controlled by limiting further expansion. This could be done by an increase in shooting pressure (at present low) in the open season, coupled with the licensing of farmers in certain areas to shoot during the close season, as advocated by Ogilvie (1977).

Acknowledgements

I am grateful to C. R. G. Campbell, G. B. Cross and K. Lane for collecting data on Wildfowl Trust refuges, to M. A. Ogilvie and A. K. M. St. Joseph for providing information on goose numbers and distribution and to J. B. Blossom for drawing Fig. 1.

References

BELLROSE, F. C. (1976) Ducks, Geese and Swans of North America. Harrisburg, U.S.A.; Stackpole Books. 544pp. ISBN 0 81117 0535 8.

BOYD, H. (1956) Statistics of the British population of the Pink-footed Goose. *Journal of Animal Ecology* 25, 253–273.

BOYD, H. (1974) Comment on a proposed reduction in the number of Barnacle Geese wintering in Islay. London, Nature Conservancy Council unpublished report, 24pp.

CABOT, D.; WEST, B. (1973) Population dynamics of Barnacle Geese *Branta leucopsis* in Ireland. *Proceedings of the Royal Irish Academy* 73 Sect. B. 415–443.

DEANS, I. R. (1979) Feeding of brent geese on cereal fields in Essex and observations of the subsequent loss of yield. *Agro-Ecosystems* 5, 283–288.

FRAME, J.; PATTON, D. L. H. (1976) The effect of winter grazing by wild geese on spring and early summer pasture production. *Journal of the British Grassland Society* 31, 41–42.

GILBERT, B. (1977) Uncle Sam says SCRAM! *Audubon* 79, 42–55.

GIVENS, L. S.; NELSON, M. C.; EKEDAHL, V. (1964) Farming for waterfowl. In: Linduska, J. (*Ed.*) Waterfowl Tomorrow. Washington; U.S. Government Printing Office p. 599–610.

JENSEN, W. I.; WILLIAMS, C. S. (1964) Botulism and Fowl Cholera. In: Linduska, J. (*Ed.*) Waterfowl Tomorrow. Washington; U.S. Government Printing Office p. 333–341.

MACINNES, C. D.; DAVIS, R. A.; JONES, R. N.; LIEFF, B. C.; PAKULAK, A. J. (1974) Reproductive efficiency of McConnell River small Canada Geese. *Journal of Wildlife Management* 38, 686–707.

OGILVIE, M. A. (1977) The numbers of Canada Geese in Britain, 1976. *Wildfowl* 28, 27–34.

OGILVIE, M. A.; BOYD, H. (1976) The numbers of Pink-footed and Greylag Geese wintering in Britain: observations 1969–75 and predictions 1976–80. *Wildfowl* 27, 63–75.

OGILVIE, M. A.; ST. JOSEPH, A. K. K. (1976) The Dark-bellied Brent Goose in Britain and Europe, 1955–76. *British Birds* 69, 442–439.

OWEN, M. (1976) Factors affecting the distribution of geese in the British Isles. *Wildfowl* 27, 143–147.

OWEN, M. (1977) The role of wildfowl refuges on agricultural land in lessening the conflict between farmers and geese in Britain. *Biological Conservation* 11, 209–222.

OWEN, M. (in press) Abdominal profile – a condition index for wild geese in the field. *Journal of Wildlife Management.*

OWEN, M.; THOMAS, G. J. (1975) Wildfowl and Agriculture. Sandy; Royal Society for the protection of Birds, 16pp.

OWEN, M.; NORDERHAUG, M. (1977) Population dynamics of Barnacle Geese *Branta leucopsis* breeding in Svalbard 1948–1976. *Ornis Scandinavica* 8, 161–174.

RAVELING, D. G. (1978) Dynamics of distribution of Canada Geese in winter. *Proceedings of the North American Wildlife and Natural Resources Conference* 43, 206–225.

REEVES, H. M.; DILL, R. H.; HAWKINS, A. S. (1968) A case study in Canada Goose management: the Mississippi Valley Population. In: Hine, R. L.; Schoenfeld, C. (*Eds.*) Canada Goose Management. Madison, U.S.A. Dunbar Educational Research Services, p. 150–165.

SANDERSON, G. C. (1976) Conservation of Waterfowl. In: Bellrose, F. Ducks, Geese and Swans of North America, Harrisburg, U.S.A.; Stackpole Books. p. 43–58. ISBN 0 8117 0535 8.

ST. JOSEPH, A. K. M. (in press) An assessment of population levels and the use of farmland in Britain by *Branta bernicla bernicla* in 1977–78. *Proceedings of the Technical Meeting on Western Palearctic Migratory Bird Management*. Paris, Dec. 1977.

Discussion

Cuthbert: I would like to introduce the possibility of physical deterrents and very briefly describe some work that I have done on Islay to inhibit grazing of barnacle geese by preventing them having free access of movement. We did this by setting up grids of trip-wires about 20 cm high of black polypropylene twine. Apart from a novel effect of about 10 days the overall effectiveness was absolutely nil. The next stage was to take up a suggestion that Janet Kear made of using electric fences. A small area 40 m square, was fenced off with an electric fence about 25 cm high. This gave very satisfactory protection for a period of about two months as judged both from droppings counts and the height of grass. I think that electric fences may have a potential for alleviating the effect of geese grazing on pasture.

Lazarus: Do we know what birds do in the short term once they have been scared from a field? Do they sit nearby and wait, or do they feed immediately somewhere else?

Inglis: Although we do not have marked populations and, therefore, our observations are to some extent suspect, we think that woodpigeons, at least, fly to nearby trees and sit there before slowly drifting back. Kenward has reported that the shorter the time pigeons have spent feeding before being disturbed the sooner they returned to that site. He argues that this reflects the level of the food reserves in their crops.

Bransden: I am very much given to over-simplification but it strikes me that over the last two days we have been hearing some very erudite lectures on bird scaring and we have got away from the basic problem of why the birds are there at all. They are there because they are hungry and if we chase them around by scaring them they will go to our neighbours. As we have to look at research money all the time, should we be looking for a cheap means of feeding them to keep them off an expensive crop rather than an expensive means of chasing them off our farm onto the next farm?

Wright: This point raised by Mrs Bransden has exercised our minds for many years. Lack taught us that food above all other factors is that which limits population size and, therefore, if you provide animals with additional food this may enable more to survive. If you provide additional food at the time when food is limiting then you will increase the total size of the population and you will be back to square one in that birds will return, looking for food, but now there will be more of them. On the other hand if the damage to the crop is known to be occurring when food is not a limiting factor then providing alternative food does, of course, offer a possible approach. It is generally believed, for example, that clover supply throughout February and March limits the woodpigeon population, so if you feed woodpigeons then you will probably have more woodpigeons the next year.

Krebs: I would like to add just one sentence to that. Although Lack emphasised the importance of food in limiting bird populations the direct evidence for it is very slim and we simply do not know what happens if you add extra food to bird populations because people have not done it, with one or two exceptions such as with the red grouse in Scotland.

Feare: I do not think that provision of some sort of alternative food source should be eliminated as a possible means of preventing damage. During the last few years, we have seen a couple of examples of this. In 1974 at Bridgets Experimental Husbandry Farm they were very late ploughing the maize stubble and all the starlings that arrived on the farm after their immigration fed there. The problem came when they eventually managed to plough the maize stubble and then the birds moved into the calf units; obviously while the maize was present it was attracting the birds out of the buildings. In a winter situation like this, when the starlings do not have a particularly high mortality, there was not a great deal of danger of increasing the population although of course you may have increased the local farm population for that winter. The other example we have is a farmer nearby who suffers bad rook damage to germinating maize and now he uses an adjacent field to scatter waste barley when the maize is at the vulnerable stage. He finds the rooks will take the barley in preference to the maize which they would have to dig up. There could be a problem here in that this was at a time when rooks might be short of food and, therefore, there could be a real danger of increasing their numbers.

Wright: Could I just mention one example of this in connection with fruit. It has been a practice with some fruit growers who have suffered from bullfinch damage to plant myrobalan hedges as bullfinches will eat them in preference to other fruit buds. However, those growers I know that have done this, have suffered no less bullfinch damage as a result.

Stanley: We have heard very recently about putting out grain for brent geese. I think this is the first case of wild brent geese feeding on grain. Are we running the risk of developing a new pest problem by introducing species to novel foods?

Owen: Yes, although I think that it is more likely that you are speeding up a natural progression rather than actually introducing novel foods. It has been suggested that pink-footed geese in Lancashire learnt to eat carrots as a result of farmers leaving heaps of them around in fields. This possibility has to be borne in mind particularly where new cash crops which might become acceptable to the birds are involved.

Rogers: To what extent is agricultural bird damage a problem of individual farmers? Put another way, is it an acceptable solution to spread the damage out among more farmers so that many farmers each lose a little bit of their crop rather than a few farmers losing it all? It seems likely that most repellents and scaring techniques are liable to result in this.

Owen: The North American experience with dispersal techniques is not very encouraging. The birds have been adapted to be gregarious and to break down that barrier I believe is going to be very difficult. I think in this country, the problem of agricultural damage by geese at least, is a very minor thing as a country-wide problem, as we saw the distribution of geese is very, very localised, and it is really only a problem in these local areas. We have not tried, as yet, to disperse the birds, but I suspect that if we did we would have very little success.

Smith: Can I ask if anyone is doing any research to try to encourage natural predators of very bad pests? For example, if you go along many of our modern roads you see lots of hawks and so on; we appear to have provided a nice feeding site for these birds. What is actually there that is attracting these birds? If we knew what was making them come to these particular situations might it be possible to encourage raptors in agricultural situations?

Krebs: Robert Kenward studied the habitat requirements of the goshawk which is a major predator of woodpigeons and one of his conclusions was that in the south of England there just is not enough woodlands for it. The sort of birds we get on motorway verges unfortunately are feeding on mice and I think are not bird predators.

Smith: Nevertheless, could they perhaps be scaring birds off adjacent arable crops, does anyone know whether cereal crops adjacent to motorways suffer less damage?

Krebs: Kenward found less damage in fields adjacent to roads, but this probably was a result of greater disturbance by man.

Feare: Kestrels frequently hover over the calf unit at Bridgets Experimental Husbandry Farm and the starlings take no notice at all, they do not even mob them.

Broom: In relation to the effectiveness of man as a bird scarer I would like to make a very general point. I feel perhaps one should consider what is likely to happen to agriculture in the long term in this country. The general trend for a number of years has been towards increasing mechanisation and a reduction in manpower, but I wonder whether this is going to continue. Perhaps we are returning to a situation where there will be more labour available, and certain aspects of agriculture will be done in a very labour-intensive way. It seems the overall result of a lot of research is that it is very difficult to deal with bird pests. Birds are very adaptable; they can modify their behaviour a great deal according to the methods which are used to try and combat them. Possibly we should look more carefully at the original method of bird control, that is to have people whose job it was to keep birds off the areas which were most sensitive to bird attack. At the moment there are in all areas of this country people who are unemployed, perhaps we should try to use such people. Looking forward into the future there are going to be more and more people wanting to go into the country and do a job which is fairly straightforward. Perhaps we should be seeking the most efficient ways of using manpower directly to frighten away birds.

Brough: It seems to me that the feasibility of this approach is largely a matter of money. If there is plenty of money available to employ these people you will employ them. A good example is the bird control situation that one has on airfields. Airfields are very expensive to maintain and, therefore, it is justifiable to have people literally running around airfields scaring birds. I would guess that agricultural crops are not quite so valuable, and that is why you do not have many people running around farms scaring the birds.

St. Joseph: I would like to emphasise the importance of prediction of possible damage as a means of crop protection. For a number of years we have found the most effective way of solving the brent goose problem is by providing advance warnings and encouraging the use of voluntary refuges. Inland feeding by brent is positively correlated with the number of young geese present in the population. We can predict this in October and as crop grazing does not really start until December or January we are able to circulate all farmers who have had geese in previous years with a letter advising them either to expect more or fewer geese than in the previous year and of appropriate scaring methods. What they then do is selectively scare brent from cereals that might be damaged and leave the geese alone on grassland. As far as we can see this is extremely effective. The number of Essex farms grazed by geese over the last three years has increased from about 50 to nearly 90, yet the number of complaints from farmers has dropped from 20 to zero. In contrast, on the south coast of England where this tactic has not yet been adopted, the number of farms being grazed by the geese has increased and so has the number of complaints.

O'Connor: I would like to point out possible ways of making predictions as to the likely impact of certain species such as bullfinch and linnet. The British Trust for Ornithology collects a variety of population dynamics data on these species, data covering such items as the number of territorial males in spring, breeding performance, their moult schedule, adult and juvenile survival, and their movement patterns. These data are available on an annual basis and their relationships can be used to predict, on the basis of long runs of past data, what the species will do in the coming year. Some of the information available is collected with too long a time lag to be useful for predictive purposes but other data do seem to offer some hope of getting a reasonable estimate of what the birds are going to do before they actually do it. From the work we have done so far with non-pest species it is clear that there are large differences between species as to the best predictor but the results are sufficiently encouraging to suggest they might be useful in the context of predicting bird pest problems before they occur.

Inglis: If I could just follow up that point of increased communication, the woodpigeon in this country is a sporting bird and I would have thought that good liaison between local shooting clubs and farmers that have woodpigeons might alleviate this particular problem.

Harradine: The Wildfowlers' Association of Great Britain and Ireland is trying to do just this, to encourage greater communication. We provide a woodpigeon and rabbit control service on a 24 hour basis. Thus, if a farmer with a pigeon problem gets in touch with his local WAGBI affiliated club, our members will go to his farm to help him in his efforts at control.

Haskell: One strategy that has not been mentioned so far is the development of resistant varieties, what about resistant variety breeding? That is my first point, and second, an old, perhaps very naive point, do we know enough about the physiology and reproduction of birds to think about a chemical means of controlling reproduction?

Jones: Trials of resistant varieties are most frequently conducted in a choice situation where you can, of course, find a difference in damage levels, but when you present birds with just a resistant variety then, in my experience, it will be damaged just as much as a non-resistant variety.

Wright: Of five possible approaches to bird damage problems that I have managed to introduce into the Joint Consultative Organisation system recently, this was the only one that was unequivocally thrown out on the grounds that it was at least a 20 year programme to develop a viable commercial variety and that at the end of that time they could find themselves with a bird-resistant variety that nobody wanted. I obviously do not think we have heard the last word on the subject but that is the scale of the problem. Frankly, there is great reluctance to take on this particular attribute of plant breeding. I know it has been done accidentally with at least one variety of cereal where some resistance associated with the glume of the seed has been found.

Drent: I would like to make one general comment about the needs of research and whether money is being well spent. We have heard something about the problems of getting acceptance by government agencies of research proposals and I think from the point of view of the biologist most programmes that are government funded are funded for too short a time. Whilst it is logical that someone who is facing a problem wants an answer very quickly, I think I take the view that if you are going to set aside a sum of money to solve that problem it is far more effective to hire, let us say three people for five years rather than 15 for one and governments always will hire 15 people for one year. I am sure that in almost all cases where there have been benefits from a research programme, you will find this was because the work was continued long enough. The group had enough continuity to allow techniques to be passed from one to another and there could be discussion about the problem rather than about the administration of the funds and the hiring of all sorts of people who will disappear before their reports are properly written up.

Haskell: I acknowledge that this has been a problem to us, but we have now rules for research which say that we must have, or expect, a return for this work in a reasonable period of time. It is a great triumph for the scientists of course to have that last phrase put in. I think it is an important point to make that, for example, if you are going to do a real study of bird physiology then five years seems to be an absolute minimum. Maybe it is a point that should be brought out in this symposium that this work does need long-term funding.

Thearle: To answer your question concerning reproduction inhibitors we have been doing some work on this but as yet have not found a sufficiently good inhibitor for use against birds. What is required is something that is long term and does not produce harmful side effects. Those inhibitors we have tested which have produced fairly long-term effects, lasting a few months anyway, also proved to be highly toxic and those which were not toxic lasted only for a matter of a few weeks so that unless you could keep on re-applying the inhibitor, which presents great baiting problems, you would get nowhere. One other point, whereas if you are using a poison or a stupefying bait and the birds are taking bait you can see that harm is being done, if you are putting down a reproduction inhibitor you do not know what harm you might be doing until it is too late. So although we are looking into this subject we do not think there is much hope at the moment.

Krebs: I hesitate to break off this discussion which has been very lively and wide-ranging, but I think that alternative food sources are now available. I would like to thank all this morning's speakers who have given excellent talks and also those people who have taken part in this discussion; thank you very much.

CHEMICAL BIRD REPELLENTS

Introduction

P. T. Haskell
Overseas Development Administration
Centre for Overseas Pest Research

Ladies and Gentlemen. We now come to the final session of the Symposium, dealing with chemical bird repellents. This is, of course, in one sense a continuation of this morning's session on bird scaring which dealt with visual, acoustic and behavioural deterrents but it justifies a session to itself both as representing a technique already widely employed – although with varying degrees of success – and one deserving of a good deal more research into basic aspects of bird physiology and behaviour. It is also, of course, probably the most widely used technique of protection in the commercial field and positive developments here could have considerable economic significance. Despite widespread interest, both amateur and professional, in bird ecology and behaviour, studies of feeding behaviour and related physiology in birds have been sadly neglected and I am sure this is something that, if corrected, would pay handsome dividends in relation to the development and utilisation of chemical repellents. If I may draw a parallel with my own discipline of entomology, progress in developing and using anti-feedant chemicals to deter insect pests has been very considerably assisted by studies on feeding behaviour and physiology, including anatomical, behavioural and neuro-physiological studies on the sensory receptors involved, how they affect choice of food, and the role of the neuro/hormone system in testing quality of food and regulating quantity ingested, and I am sure that similar studies on birds would greatly assist progress in this field.

In this session, therefore, the organisers decided it would be desirable to begin with a paper reviewing the background to the development and use of chemical repellents so as to underline the urgent need for more research and more information in this area. There can be no one better fitted to do this than Ernest Wright, who has probably more experience in this field than anyone else in the UK, and it therefore gives me great pleasure to call on him to present his paper, "chemical bird repellents – a review".

Chemical repellents – a review

E. N. Wright
Ministry of Agriculture, Fisheries and Food
Worplesdon Laboratory

Summary

Chemical repellents have probably been used by man ever since he first began to cultivate crops. Understandably his choice of materials was anthropocentric but, despite the fact that birds and mammals have similar sensory systems, birds do not appear to regulate their food intake by reference to the same cues as man. The sense of taste, in particular, might be expected to play a central role in discriminating between what is and is not acceptable as food, but this function has yet to be convincingly demonstrated in birds. Materials having an unpleasant e.g., bitter, flavour simply do not deter birds and though tactile e.g., sticky, substances have more effect they suffer serious practical limitations for use as repellents. It is clear that birds, and other animals, can learn to avoid food that makes them ill and the inducement of such "conditioned" aversion is one possible approach to the development of effective repellents. In the long term, an understanding of the mechanisms by which animals select their food could lead to new concepts in the prevention of damage by pests.

Introduction

It is probable that man has used some kinds of chemical repellents to protect his crops since the earliest times. Plants with pungent odours, such as the onion family, and those with an unpleasant taste, like quassia, feature in countless cottage remedies for driving away pests. That people should attempt to use such substances in order to repel animals is understandable – we find them unpleasant, therefore, other species having similar sensory systems will find them unpleasant – but the origin of other alleged repellents is more obscure. For instance, the Caper Spurge (*Euphorbia lathrys*) is widely held to repel moles (*Talpa europea*) and old gardening books, and even some modern ones, frequently recommend its cultivation as a way of ensuring a mole-free vegetable patch. Gervase Markham (1636) in his book "Inrichment of the Weald of Kent" suggests that moles will not come within 60 yards of a Caper Spurge plant in its second year of growth. That this observation is false I can testify from personal experience; yet somehow the myth survives.

The folklore of repellents is not only concerned with plants and plant extracts; foul-smelling animal oils, such as bone oil and some fish oils, faeces and urine all feature as ingredients in traditional repellent mixtures. Perhaps the most ancient repellent of all is lime manure, prepared by mixing five parts lime, four parts cow dung and twelve parts pig dung with just sufficient water to give an even, free-flowing mixture; it was used mainly on fruit trees and bushes against bud-eating birds. Whilst humans might generally find such concoctions repulsive the reaction of other species is by no means so predictable. Although birds possess the same basic sensory mechanisms as mammals it would be naive in the extreme to assume that their perception of stimuli corresponds to our own and, even within the class Aves, considerable variation is to be found in the degree of development and functional rôle of the specialised sensory organs. Since an aversive response to a repellent is first dependent upon the animal perceiving the stimulus, when considering bird repellents it is pertinent also to consider the sensory mechanisms of birds.

The chemical senses

Chemical stimuli are mainly perceived through the specialised senses of smell and taste and although functionally it is very difficult to separate one from the other it is convenient to consider them as independent attributes.

Olfaction

The sense of smell in birds has been the subject of controversy and argument for the past century and a half, in fact ever since Darwin (1834) and Audubon (1826) addressed themselves to the question of whether birds can smell. Although anatomists have frequently drawn attention to the well-developed olfactory organs of some birds, direct evidence of olfactory behaviour has been wanting and the view that birds are largely anosmatic has tended to persist in the literature. During the past 25 years, however, several workers (Michelsen 1959, Stager 1964, 1967, Tucker 1965, Henton *et al.* 1966, Wenzel 1968, Grubb 1971, 1972, Papi 1971, 1972, Hartwick 1977) have convincingly demonstrated the ability of birds to smell. Stager (1967) sums up the position by saying "Compared with the demonstrated macrosmatism of certain insects and mammals, it can be safely stated that the class Aves on the whole is microsmatic, but to accept the belief that birds are anosmatic is comparable to arguing that the class Mammalia is unable to fly . . . ". It is now recognised that species of birds with a highly developed sense of smell are to be found in several avian Orders e.g., APTERYGIFORMES: kiwis (*Apteryx* spp.), FALCONIFORMES: the turkey vulture (*Cathartes aura*), PROCELLARIIFORMES: Leach's storm-petrel (*Oceanodroma leucorrhoa*), CAPRIMULGIFORMES: the oil bird (*Steatornis caripensis*) and PICIFORMES: the honeyguides (*Indicator* spp.). All of these birds seem to depend, at least to some extent, on their sense of smell to find food; the colonial nesting species probably also employ smell as a homing aid and for individual recognition. Even more recent studies (Hartwick 1977, Papi 1972) indicate that smell may be important in the homing behaviour of pigeons – a species shown to be poorly endowed with olfactory epithelia. In summary, it seems fair to conclude that the olfactory senses are functional in most birds but only a few species exhibit odour-related behaviour. This evidence does little to encourage belief in the repellency of odours. If most birds disregard olfactory clues then the question of whether they actually perceive odours is academic since in practice it will prove impossible to manipulate their behaviour via the olfactory channel of communication. Clearly this is not true for those species that have been shown to seek food by smell; in their case false trails could be laid which might totally disrupt normal feeding behaviour. It might also be possible to obscure the smell of food by introducing a screening odour, as is done in many so-called "air freshners". But the species known to have an acute sense of smell are not pests and, unless future research reveals otherwise, we must regard birds that eat our crops as microsmatic. As has already been pointed out, this does not mean that they cannot smell and it would be surprising if they did not show an initial response to intense olfactory stimulation. In man some olfactory stimuli are so intense that they surpass the threshold of pain; ammonia readily produces such an effect and this has been used, with some success, against roosting birds. However, questions of cost and practicality apart, the release into the environment of intense odour-producing substances is potentially hazardous and unlikely to be tolerated by a pollution conscious society.

Gustation

That birds have a functional sense of taste is a well established fact but their degree of sensitivity is less well-defined and the role played by taste in food selection remains obscure. As in mammals, the sensory receptors for the perception of taste are the "taste buds", consisting of ovoid clusters of cells lying in cavities in the stratified epithelium of the tongue. Stimulation of the taste buds results in nervous impulses being transmitted to the brain by the glossopharyngeal nerve and, using appropriate electro-physiological techniques, these impulses can be monitored whilst the tongue is irrigated with solutions representing different taste stimuli. In this way it has been possible to demonstrate, for instance, that salts and acids generally elicit neural responses; that the response of birds to substances that taste sweet to man is minimal and whilst chickens respond to quinine, pigeons do not (Wenzel 1973). Behavioural confirmation that birds perceive taste has been sought in preference tests and a variety of different experiments all lend support to the view that birds are capable of distinguishing certain flavours (Kare *et al* 1957, Kare & Ficken, 1963, Gentle 1972). Similarly, preference tests give some indication of thresholds of taste perception and, although pigeons were able to discriminate between certain substances that were tasteless to man (Duncan 1963), in most cases the taste acuity of birds appears to be less than that of mammals. This is not surprising in view of the fact that mammals possess many more taste buds than birds; most birds so far investigated possess less than a 100 taste buds whilst the maximum number found in any species, a parrot, is 400 (Portmann 1961). By contrast it is common to find many thousands in mammals e.g., c. 9000 in man, 17 000 in the rabbit and as many as 40 000 in the cow. Nevertheless, it has been pointed out that one must not confuse proliferation of the taste apparatus with the quality of the experience and the possibility must be admitted that the sense of taste in birds may be no less acute than in mammals.

In spite of the accumulated evidence that birds perceive flavours we still have no evidence of how, or if, this influences their normal behaviour. It is logical to assume that an animal possessing a functional sense of taste would utilise that sense in the selection of food, yet with birds there is little evidence that this is the case. As long ago as 1940 Englemann concluded that hens select grains on the basis of shape and, to a lesser extent, on colour; taste playing little or no part. Kare (1961) used chickens to test the hypothesis that taste is a mechanism for the maintenance of adequate nutrition, but obtained conflicting results. He offered nutritionally deficient birds a choice between adequate and inadequate diets and, although he found some evidence that birds selected diets that would correct the deficiency, some birds preferred the nutritionally less desirable food. This cannot be taken as convincing evidence that birds do not regulate their nutrition through taste since other workers have shown that induced thiamine deficiency results in an increased intake of thiamine-supplemented diet (Hughes & Wood-Gush 1971).

Numerous investigators have attempted to influence the acceptability of food to birds by adding flavours and this is essentially the basis of most work with chemical repellents. Indeed much of the work on the sense of taste in birds stems directly or indirectly from pursuit of this practical goal. However, in this respect, as with olfaction, the experimental evidence is not encouraging. In his preference tests Duncan (1960) discovered that pigeons would not drink water containing even very low concentrations of quinine and other workers (Brindley 1965, Kare *et al*, 1957, Cane & Vince 1968) also demonstrated a rejection of bitter substances by a variety of species. At this stage the experimental evidence appeared to uphold the traditional reputation of quinine as a bird repellent but when hungry birds were presented with food treated with quinine, and other more intensely bitter substances, it was readily eaten (Wright, unpublished). Although further work with chickens has shown that with increasing concentration of the aversive flavour total food consumption may be reduced (Kare 1961), this effect has not been satisfactorily demonstrated with wild birds in the field.

In all this work great variation is to be found between species, between individuals and in the reaction of the same individuals to chemically similar substances. At present we have little idea of the factors leading to acceptance or rejection of a compound by birds and much further research is required, particularly in the behavioural context. An interesting step in this direction is the relatively recent discovery of taste-active proteins (Cagan 1973) which are thought to hold considerable promise as experimental tools in taste research. The intensity ($10^5\times$ more effective than sucrose) and persistence of the sweet taste of monellin makes this substance particularly interesting in view of the variable responses shown by birds in preference tests with sugars. Miraculin, the taste-modifier protein, when applied to the tongue, causes acids to taste sweet and this might prove especially useful in behavioural research.

For the present, at least, we must accept the evidence that taste *per se* is not very important to birds and this suggests that unpleasant, but harmless, flavours are unlikely to prove repellent. But taste stimuli are rarely encountered in isolation; certainly the ingestion of food represents a complex sensory experience involving visceral, tactile, olfactory and thermal information in additon to that concerned with flavour, hence behaviour that gives every appearance of being taste-orientated may be controlled by other factors.

Other senses

The remaining senses, *auditory, visual* and *tactile* – are not primarily concerned with chemo-reception but chemicals can be employed in ways that affect both appearance and texture and thus chemical repellents might be designed to act in this way. In the visual field, the only aspect I shall consider is that of colour perception, for which there can be no serious doubt in the case of birds. Many species are brightly coloured and in territorial or sexual displays these features of the plumage are often accentuated by posturing. It can be shown experimentally that a tuft of appropriately coloured feathers, or even a coloured disc, can function as a releaser, triggering off a whole chain of behavioural responses (Lack 1939, Tinbergen 1953). Given sufficient understanding of the underlying behavioural patterns for individual species one can postulate that subtle use might be made of such mechanisms; at least during the breeding season. At a more practical level, colour can be used to change the appearance of food and this might have a repellent effect through some innate abhorrence for the colour used or, more probably, cause the food to be overlooked on account of the chromatic disguise. There is ample experimental evidence to support the view that birds perceive about the same range of colours as man (Pumphrey 1961) and some indication that their vision is more sensitive at either end of the spectrum than in the yellow-green region (Dücker & Schulze 1973). If there is any truth at all in the view that birds avoid red-coloured objects it must surely be because they appear conspicuous, which implies a certain element of novelty which we know influences behaviour (see Inglis 1980).

Tactile stimuli may be of many kinds but for convenience the receptor organs are often considered to be of three types; those concerned with pressure, with temperature and with pain. Physiologically little seems to be known about these sensory systems. The skin and internal organs are rich in free nerve endings that terminate in a variety of structures that function as receptors and the buccal cavity, especially the tongue, of birds is well endowed with corpuscles of Merkel which are believed to have a tactile function. This would accord with the view that birds are extremely sensitive to the surface texture of food (Englemann 1943) and when they are seen to mandibulate unusual items of diet there is good reason to suppose they are exploring the feel of it rather than its flavour (Ligon & Martin 1974). Many years ago I observed that the bullfinch (*Pyrrhula pyrrhula*), which feeds on buds throughout the winter, will not eat buds covered with hoar frost; this may, of course, be due either to the fact that they are no longer recognised as food or because they are very cold, but dislike of the textural change is a plausible alternative explanation. Further evidence supporting the belief that texture is more important than flavour comes from field experiments with repellent formulations; coarse stone dust, with a suitable sticker, was more effective than chemical compounds having strong odours and flavours in deterring bullfinches from eating fruit buds (B. Jones, personal communication). Sticky substances, which indisputably impart strong tactile stimuli, are perhaps the most effective existing bird repellents, but by their nature they are suitable for only a limited range of applications. Originally natural gums were used but nowadays chemicals belonging to the polyethylene or polybutene groups are more often employed. Where the problem is to repel birds from specific roosting places, such as window ledges and roof girders, these materials can be very effective but they are obviously quite unsuitable for application to ripe or ripening crops – a clear case of the cure being worse than the disease.

Although chemical reactions can be used to produce heat changes it is difficult to foresee any practical application of such techniques to repel birds. It is nevertheless interesting to note that heat screens, both radiant and hot-air, have been tried as a way of deterring birds from entering warehouses through loading-bays and other access points that are necessarily kept open. Conversely, in the belief that minor differences in surface temperature determine where ground-roosting birds congregate, at least one attempt has been made to manipulate the roosting behaviour of gulls on an airfield. By installing subsoil heating in a remote, corner of the airfield, it was hoped that gulls would be attracted to roost in this "comfort zone" rather than on the runway. The apparent lack of success may be attributed to the small scale of the experiment and failure to take into account other factors that influence choice of roosting site.

Pain stimuli are almost certainly important as the sensory mechanism involved in aversive conditioning; this is typically seen as a response to particular food following sub-lethal poisoning. It is not without significance that many pesticides, especially the more toxic insecticides, have acquired a reputation for repellency to birds. In some cases, where bird damage to a crop has ceased following application of a pesticide, this should undoubtedly be attributed to the death of the birds, but there are now many well-substantiated examples in which mortality does not appear to be the major factor. It is interesting that conditioned aversion occurs more readily in connection with certain poisons than with others. For instance, in Britain a licence is required to use bait containing alpha-chloralose to control harmful birds and it is a condition of the licence that any protected species that are accidentally taken shall be cared for until they recover and then be released. If such resuscitated birds are released while the treatment is still in progress they frequently return immediately to the bait and ingest a further dose; indeed "bait happy" birds can become quite an embarrassment to operators. By contrast, certain carbamate compounds seem to possess properties conducive to the establishment of an aversive association which can result in a dramatic avoidance of further contact with the chemical. Methiocarb was the first of this group of compounds to be identified as possessing properties that made it potentially useful as a bird repellent and, despite its relatively high toxicity (LD_{50} to the starling of about 15 mg/kg), its use as a repellent on certain crops has now been cleared by the United States Environmental Protection Agency. Experimental work with methiocarb used as a repellent, has been on a very small scale in Britain and the results have not been very promising. Perhaps the small scale of the experiments has militated against success but it is right to be cautious with compounds that have a narrow margin between the lethal and repellent dose levels. One might even question whether very toxic compounds can qualify to be called "repellents".

One answer to this problem of toxicity might be to try to produce repellents that simulate natural systems evolved by insects as chemical defence mechanisms against vertebrate predators. Most of the systems so far investigated involve plants that manufacture compounds that are poisonous to vertebrates but not to insects that feed on the plant. Consequently the insects ingest the poison and themselves become toxic to vertebrate predators. The poisons involved in these systems are mostly cardiac glycosides which have a specific action on the vertebrate heart but they also have side effects, one of which is to stimulate vomiting. It has been found experimentally that the emetic dose is about half the lethal dose, thus vomiting acts as a safety factor which operates to rid the animal of the poison and normally prevents retention of a fatal dose. As Brower (1969) has pointed out, in a system of this kind there are three levels at which an insectivorous bird may reject a poisonous insect. Firstly, basic gastronomic rejection brought about by the effect of the poison; this is inefficient because the bird becomes ill and loses any food that may already have been in its crop. Subsequently, as a result of such an experience, the bird will probably learn to reject food of that particular flavour, hence further illness can be avoided but only after the prey has been chased and caught; this too is wasteful. The most efficient mechanism of rejection is when the bird associates unpalatability with the visual characteristics of the food. It has been suggested that aversive conditioning could be applied to the solution of certain mammalian predator-prey problems (Gustavson *et al* 1974), where it is obviously essential for rejection to occur at the visual level, and similar thinking has been extended to large raptors (Brett *et al* 1976). Although practical application of these techniques presents many problems, and confirmation of field success is still awaited, these experiments are a stimulating development in repellent research.

Behavioural factors

Another mechanism of repellency that must be considered is that of behaviour-modifying chemicals; for instance, everyone is now familiar with insect pheromones and their powerful species-specific effects. Unfortunately there is no evidence for the existence of any similar mechanism in birds, although a few species, as was pointed out earlier, appear to use olfactory cues both in finding food and homing. Some of these birds have characteristic body odours, which probably aid in mate recognition but, as yet, nobody has been able to link specific behaviour patterns with particular odours. However, chemicals have been used to modify behaviour in a different way. The chemical 2,4 amino-pyridine, when ingested by birds, apparently causes great panic, possibly resulting from intense pain, combined with a degree of unilateral paralysis. Birds that have eaten baits treated with this chemical tend to utter distress cries and to fly in spirals before exhibiting total collapse and death. Conspecifics hearing the distress calls, and observing the abnormal behaviour, usually flee. Thus, for gregarious species, feeding flocks comprising many thousands of birds may be scared away by the behaviour of a few individuals. The technique employed is to scatter a bait containing 1 or 2% of treated particles; in this way sufficient bait can be laid to prove attractive whilst ensuring that relatively few birds will be affected. Many people consider this method of bird control to be inhumane, and its use is prohibited in some countries, but there can be no doubt that it works. Perhaps it is not too much to hope that other chemicals will be found which can influence the behaviour of birds in more subtle and less painful ways.

In any consideration of animal repellents the question of how wild animals select their food is central to the problem. The feeding behaviour of all species has evolved as a way of maximising the utilisation of resources whilst keeping inter and intra specific competition to a minimum. As a result, different species mostly eat different foods but the diet of each contains a range of items, the relative proportions of which change markedly with time. To some extent these changes reflect availability – starlings would be hard pressed to find cherries in mid-winter – but they are also believed to be determined by nutritional requirements. This implies a mechanism of selection to meet bodily demands and this could operate either by positive cues, which would stimulate the animal to eat particular food items or negative ones having exactly the opposite effect. Clearly both systems operate in man, through several sensory channels, and we consciously respond to the smell, sight, texture and flavour of food with pleasure or revulsion. To my mind, the fact that birds do not respond to the basic taste modalities in the same manner as humans does not rule out the possibility that their feeding behaviour is governed by a similar mechanism. Indeed, it is becoming clear that the minor chemical constituents of plants play an important role in determining their acceptability as animal food. If we study these systems in action we may discover not only the key to the development of effective bird repellents but an entirely new concept in the way to prevent damage by pests.

References

AUDUBON, J. J. (1826) Account of the habits of the turkey buzzard (*Vultur aura*) particularly with the view of exploding the opinion generally entertained of its extraordinary power of smelling. *Edinburgh New Philosphical Journal* 2, 172–184.

BRINDLEY, L. D. (1965) Taste discrimination in Bobwhite and Japanese quail. *Animal Behaviour* 13, 507–512.

BRETT, L. P.; HAWKINS, W. G.; CAREIN, J. (1976) Prey-Lithium aversions III, Buteo hawks. *Behavioural Biology* 17, 87–98.

BROWER, L. P. (1969) Ecological Chemistry. *Scientific American* 220 (2), 22–29.

CAGAN, R. H. (1973) Chemostimulatory Protein: A New Type of Taste Stimulus. *Science* 181 (4094), 32–35.

CANE, V. R.; VINCE, M.A. (1968) Age and learning in quail. *British Journal of Psychology* 59, 37–46.

DARWIN, C. (1834) Naturalists' Voyage around the World. New York, D. Appleton & Co.

DÜCKER, G.; SCHULZE, I. (1973) Das Farbunterscheidungsvermoyen bei *Sturnus vulgaris* (Sturnidae) und *Padda oryzivora* Estrildidae). *Zeitschrift für Tierpsychologie* 32, 496–505.

DUNCAN, C. J. (1960) Preference tests and the sense of taste in the feral pigeon. *Animal Behaviour* 8, 54–60.

DUNCAN, C. J. (1963) The response of the feral pigeon when offered the active ingredients of commercial repellents in solution. *Annals of Applied Biology* 51, 127–134.

ENGLEMANN, C. (1940) Versuche über die Beliebtheit einiger Getreidearten beim Huhn. *Zeitschrift für vergleichende Physiologie* 27, 525–544.

ENGLEMANN, C. (1943) Uber den Geschmacksinn des Huhnes. VIII. Der Einfluss zusätzlichen Geschmacks auf die Annahmehaufigkeit fester Futterstaffe durch Zwerghühner. *Zeitschrift für Tierpsychologie* 5, 552–574.

GENTLE, M. J. (1972) Taste preference in the chicken *Gallus domesticus* L. *British Poultry Science* 13, 141–155.

GRUBB, T. C. (1971) Olfactory navigation by Leach's Petrel and other procellariform birds. Doctoral Dissertation. University of Wisconsin, Madison.

GRUBB, T. C. (1972) Smell and foraging in Shearwaters and Petrels. *Nature* 237, 404–405.

GUSTAVSON, C.; GARCIA, J.; HAWKINS, W.; RUSINIAK, K. (1974) Coyote predation control by aversive conditioning. *Science* 184, 581–583.

HARTWICK, R. F.; FOA, A.; PAPI, F. (1977) The effect of olfactory deprivation by nasal tubes on homing behaviour in pigeons. *Behavioural Ecology and Sociobiology* 2, 81–89.

HENTON, W. W.; SMITH, J. C.; TUCKER, D. (1966) Odor discrimination in pigeons. *Science* 153, 1138–1139.

HUGHES, B. O.; WOOD-GUSH, D. G. M. (1971) Investigations into specific appetites for sodium and thiamine in domestic fowls. *Physiology & Behaviour* 6, 331–339.

INGLIS, I. R. (1980) Visual Bird Scarers: An Ethological Approach. In: Wright, E. N.; Inglis, I. R.; Feare, C. J. (*Eds*) Bird Problems in Agriculture. London; British Crop Protection Council, Symposia series monograph 23.

KARE, M. R.; BLACK, R.; ALLISON, E. G. (1957) The sense of taste in the fowl. *Poultry Science* 36, 129–138.

KARE, M. R. (1961) Comparative aspects of taste. In: Kare, M. R.; Halpern, B. P. (*Eds*) The Physiological and Behavioural Aspects of Taste. Chicago; University of Chicago Press. p. 6–15.

KARE, M. R.; FICKEN, M. S. (1963) Comparative Studies on the Sense of Taste. In: Zotterman, Y. (*Ed*) Olfaction and Taste. New York; Macmillan. p. 285–297.

LACK, D. (1939) The behaviour of the robin: I & II *Proceedings of the Zoological Society of London.* A109, 169–178.

LIGON, J. D.; MARTIN, D. J. (1974) Pinon seed assessment by the Pinon Jay, *Gymnorhinus cyanocephalus*. *Animal Behaviour* 22, 421–429.

MARKHAM, G. (1636) Inrichment of the Weald of Kent, or a direction to the husbandman for the true ordering and manuring and inriching of all the ground within the Weald of Kent and Sussex. London; J. Harison. 24 pp.

MICHELSEN, W. J. (1959) Procedure for studying olfactory discrimination in pigeons. *Science* 130, 630–631.

PAPI, F.; FIORE, L.; FIASCHI, V.; BENVENUTI, S. (1971) The influence of olfactory nerve section on the homing capability of carrier pigeons. *Monitore Zoologica Italiano Firenze* [N.S.] 5, 265–267.

PAPI, F.; FIORE, L.; FIASCHI, V.; BENVENUTI, S. (1972) Olfaction and homing in pigeons. *Monitore Zoologica Italiano Firenze* [N.S.] 6. 85–95.

PORTMANN, A. (1961) Sensory Organs: Skin, Taste and Olfaction. In: Marshall, A. J. (*Ed*) Biology and Comparative Physiology of Birds. New York & London; Academic Press. vol 2, 37–48.

PUMPHREY, R. J. (1961) Sensory Organs: Vision. In: Marshall, A. J. (*Ed*) Biology and Comparative Physiology of Birds. New York & London; Academic Press. vol 2, 55–68.

STAGER, K. E. (1964) The role of olfaction in food location by the Turkey Vulture, *Cathartes aura. Contributions in Science. Los Angeles County Museum* No. 81, 1–63.

STAGER, K. E. (1967) Avian olfaction. *American Zoologist* 7, 415–419.

TINBERGEN, N. (1953) The Herring Gulls' World. London: Collins.

TUCKER, D. (1965) Electrophysiological evidence for olfactory function in birds. *Nature* 207, 34–36.

WENZEL, B. M. (1968) The olfactory prowess of the kiwi. *Nature* 220, 1133–1134.

WENZEL, B. M. (1973) Chemoreception. In: Farner, D. S.; King, J. R. (*Eds*) Avian Biology. New York & London; Academic Press. vol 3, 389–411. ISBN 0 12 249403 2.

Conditioned taste aversion: its role in bird damage control

John G. Rogers, Jr.
U.S. Fish and Wildlife Service
Washington, D.C.

Summary

The concept of using a repellent to manipulate the feeding activities of depredating birds is an old one. However, systematic investigations of avian feeding behaviour and how it is regulated have not frequently been conducted in support of applications of this concept. Evidence from research on mammals (rodents) indicates that the conditioned aversion is an extremely powerful response that controls feeding on toxic materials, placed in foods intentionally or, occurring in plants as secondary substances. This paper reviews some laboratory and field experiments with economically important avian species indicating that the conditioned aversion is the behavioural response to be exploited in developing avian repellents. Also reviewed are experiments demonstrating that Mesurol[R] is a repellent that produces this response and is an effective repellent in the field.

Introduction

It is obvious that the present symposium was assembled because birds damage crops to obtain food. It seems, then, most efficient that in attempting to resolve conflicts between man and birds in agricultural situations we should focus on, and attempt to manipulate, the feeding behaviour of offending avian species. If we can alter this behaviour on a long-term basis, we can solve the problem. Frequently, the manipulation of feeding behaviour is attempted through the use of "repellents" that act through the taste system.

The early literature on repellents has been summarised many times (e.g., Welch 1954, Besser & Welch 1959, Armour 1963, Rogers 1978a). These reviewers give the impression that no repellent has been developed that is consistently effective for use against birds. The reasons for this are complex, but stem principally from the fact that we have tended to be anthropomorphic in the most basic assumption of repellent action. Most humans realise that when offered an array of foods, some of which taste bad, we alter our feeding pattern to consume only those that taste good. This basic human experience has been translated directly to avian populations. The original assumption in repellent development seemed to be that if a potential food (the protected crop) could be made bad-tasting enough, the pest bird would stop eating it. That this has not often been the case is another reason for this symposium.

The preceding paper has presented a discussion of the gustatory capabilities of birds. It is important to understand the avian sense of taste because we continue to emphasise this system in attempts to develop chemical means of dissuading birds from consuming agricultural crops. That birds differ in their taste preferences as individuals, strains, or species has obvious ecological advantages. For example, it permits a population to utilise much more of the food in an environment than would be possible if all avian species competed for a limited group of foods, and it contributes to an adaptive plasticity of food habits, making the invasion of new habitats and utilisation of new foods possible. Variation in response to taste is further compounded by possible seasonal changes in sensitivity and acuity. It is interesting to speculate on whether taste directs or follows the abrupt changes in feeding pattern of birds that are insectivorous for part of the year and graminivorous for the rest. Taste might also play a role in the intensive feeding before migration.

In this paper I firstly discuss some of my thoughts and research directed at the development of repellents for birds. I then discuss some laboratory experiments that have led us to believe that the conditioned aversion is the response that is most effective in altering avian food habits on a long-term basis, and thus is the approach that should be exploited in the development of repellents. Finally, I will discuss some laboratory, controlled field, and large-scale field tests with Mesurol[R](*) which demonstrate that this is indeed the correct approach.

Conditioned aversions

The necessary behavioural response to a repellent has received little attention in the literature on repellents. On the surface it seems obvious that the required response is a cessation of feeding on the protected food source; the physiological-psychological mechanism to achieve this result is less obvious. Most vertebrates have developed a behavioural mechanism with which to cope with toxic substances in their food. A brief reading of the vast literature concerning the use of toxicants against rats demonstrates this phenomenon.

The major problem in the use of toxicants against commensal rodents is that those animals receiving sublethal doses of the poison refuse to consume any more of the bait (Rzoska 1953). This behavioural response, a problem in lethal control, is exactly the behaviour we are trying to induce with the use of repellents. This adaptive behaviour (bait shyness) has been exhaustively studied as conditioned taste aversion by psychologists. Rozin & Kalat (1971) have summarised much of the literature, principally for the rat, and have described (p. 478) how the rat handles the problem of food selection: "A rat becoming sick at a garbage dump (where he was poisoned) . . . may have eaten a few different foods. He knows it was a food that made him sick and can discount any familiar safe foods. With the capability of forming associations over long delays, he is now likely to associate his illness with the last relevant thing or few things he ate over the last few hours". Thus, the taste of a food that made an animal sick is subsequently avoided; a conditioned aversion is formed.

An additional powerful line of evidence led us to pursue the possibility of using the conditioned aversion in developing avian repellents. This evidence is that plants through evolution have emphasised chemical defences that adversely affect the physiology of their vertebrate predators (Rogers 1978a); they have not frequently used taste stimuli except as cues to toxic events (Alcock 1970).

Though it seems probable that most if not all pest species are capable of learning aversions in the laboratory, the conditions under which this aversion most readily occurs are not necessarily present in crop depredation situations. The pest is required to form an aversion to a familiar food (the crop) in combination with an aversive agent. We might then expect the learning of the aversion to be more difficult in light of the demonstrated importance of novelty in taste aversion learning (Rozin & Kalat 1971). The possible decrease in response because the treated food is familiar and has previously been

*Use of trade names does not imply endorsement by the U.S. Government.

considered as "safe" might be somewhat ameliorated by the fact that the aversive agent (repellent) would not usually be present in one food item in a vast array. It would be expected to add a novel or unfamiliar taste to a familiar food. A great proportion of crop depredations occur in near-monoculture situations (e.g., corn, rice, large fruit orchards) where, before the onset of the adverse post-ingestinal effect, the target species would be expected to have consumed only a very limited number of foods, and possibly only the crop with the repellent material added. Thus, the problem of associating the illness with a particular food would be simplified.

Mammalian species seem to be able to form conditioned aversions most readily to gustatory cues and rely upon other food-related cues (visual, olfactory) to a lesser extent (Rozin & Kalat 1971). Evidence indicates that at least one bird, the bobwhite quail (*Colinus virginianus*), may form aversions more readily to visual than to gustatory cues (Wilcoxon *et al* 1971). Nevertheless, bobwhites, and probably other avian species, were able to utilise gustatory stimuli as cues to adverse post-ingestional effects.

The first questions that concerned us were (1) could a particular pest bird species, the red-winged blackbird (*Agelaius phoeniceus*), form a conditioned aversion to a toxicant-food combination, and (2) how would the effectiveness of the conditioned response compare with that to a material that simply tasted bad and had no toxic properties. At that time we were also interested in discovering whether the candidate repellent, Mesurol[R] (3,5-dimethyl-4-[methylthio] phenol methylcarbamate), depended on simple taste or a conditioning effect for its mode of action.

In an attempt to answer these questions several materials were compared in three, two-choice tests where the palatability of the alternative to the treated food ranged from highly palatable through mildly unpleasant to highly offensive (Rogers 1974). By measuring the time the birds took to transfer feeding to the alternative it was possible to examine the motivating strength of the various repellent stimuli (Table 1). In these tests it was determined that the only materials consistently effective in altering the feeding behaviour were those that confronted the bird with a choice between illness as a penalty for continuing to eat the treated food, and the alternative bad-tasting but toxicologically harmless diet. The data from this experiment demonstrated that red-winged blackbirds

Table 1

Average number of hours required by male red-winged blackbirds to learn to avoid several repellents.[1] *Numbers in parentheses refer to numbers of birds of 18 tested that responded (sucrose octaacetate) or survived (methiocarb, LiCi). NR, no response (from Rogers 1974).*

	Compound		
Choices	SOA	LiCi	Methiocarb
Treated vs. untreated mash	2·8 (18)	2·4 (18)	2·7 (18)
Preferred vs. nonpreferred (corn or rice)	3·2 (18)	6·6 (13)	10·0 (12)
Mash vs. DMA[2] checkerettes	NR (0)	9·1 (12)	11·4 (11)

[1] Means not underlined by the same line are significantly different from each other by Duncan's New Multiple Range Test (P 0·01).

[2] Dimethylanthranilate; toxicologically harmless but highly avoided by red-winged blackbirds.

could form conditioned aversions and supported the very important contention of Alcock (1970) who suggested that, except as signals to toxic or emetic effects, negative gustatory cues from prey are of little significance in determining food preference of wild birds.

This initial experiment led us to examine further the characteristics of Mesurol^{R} and the conditioned aversion it produces in pest birds that contact it through their food. The first set of experiments toward this end were conducted in the laboratory (Rogers 1978b) with Mesurol^{R} as a model, and were designed to elucidate the time course of development of an aversion, the duration of an aversion once it had been learned, and the relative contribution of the sensory qualities of the food and the aversive agent to the aversion.

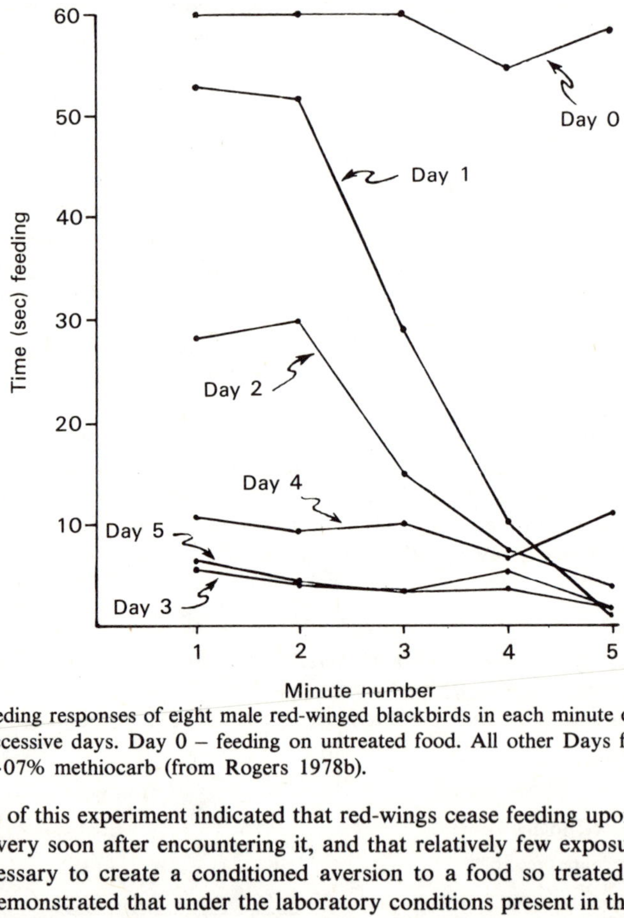

Figure 1. Feeding responses of eight male red-winged blackbirds in each minute of 5-min testing periods on successive days. Day 0 – feeding on untreated food. All other Days feeding on food treated with 0·07% methiocarb (from Rogers 1978b).

The results of this experiment indicated that red-wings cease feeding upon methiocarb-treated food very soon after encountering it, and that relatively few exposures to treated food are necessary to create a conditioned aversion to a food so treated (Fig. 1). The results also demonstrated that under the laboratory conditions present in this experiment, red-winged blackbirds could retain the aversion, without apparent extinction, after a 16 week interval (Fig. 2). The sensory properties of methiocarb also appeared responsible for the aversion in the final series of tests (Table 2). The birds were not reluctant to consume untreated food of the kind previously treated yet refused a new food treated with methiocarb.

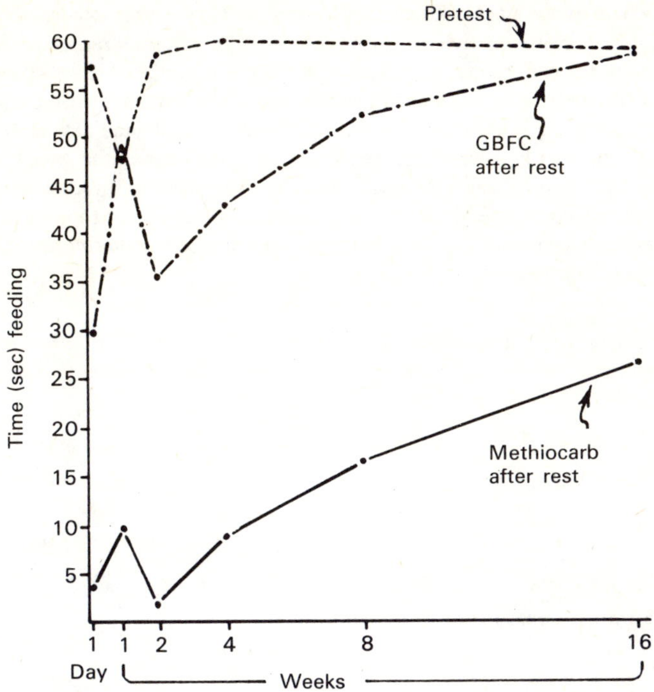

Figure 2. Feeding responses of 6 groups of 9 male red-winged blackbirds at various intervals after formation of a conditioned aversion to 0·07% Mesurol[R]. Top curve represents feeding on untreated food before training. Middle curve represents feeding on untreated food after the rest interval, the day preceeding re-testing with Mesurol[R]. Bottom curve represents feeding on treated food after the rest interval (from Rogers 1978b). GBFC – Purina Game Bird Flight Conditioner.

Table 2

Feeding response in a 1-min exposure of eight male red-winged blackbirds to untreated foods and foods treated with 0·07% methiocarb. The treatments are arranged in order of presentation from top to bottom[1] (from Rogers 1978b).

Treatment	Time spent feeding (s±SEM)
Pretest untreated GBFC[2]	57·9±2·1*†
First exposure methiocarb in GBFC	59·5±0·5*
Methiocarb in GBFC after training	5·1±3·4
Untreated GBFC	45·5±7·0†
Untreated rice	48·5±7·0*†
First exposure methiocarb in rice	17·4±3·9

[1] All means not marked with the same symbol are significantly different from each other (Duncan's New Multiple Range Test) (P = 0·05).
[2] GBFC – Purina Game Flight Conditioner.

In a set of field experiments (Rogers & Linehan 1977) we attempted to identify possible alterations in the feeding behaviour of common grackles (*Quiscalus quiscula*) feeding on newly-planted corn seed treated with Mesurol^R. To accomplish this we observed grackles feeding in a field, various sections of which had been planted with both treated and untreated corn. Among other things, this experiment demonstrated (Table 3) that grackles did not behave differently towards treated seeds, in terms of the time it took to consume a seed or the proportion of seeds dropped, whereas they did consume fewer treated seeds and did not spend more time in the untreated area. This suggested to us that they avoided feeding on treated seeds and seedlings but were not averse to entering the area where the food source was treated.

Table 3
Feeding activities of grackles that entered both the treated and untreated sections of the experimental corn seed planting.[a] *Number of birds in parentheses (from Rogers & Linehan 1977).*

	Treated	Untreated	Combined
Seconds/row[b]	5·6±0·9	7·1±1·0	6·3±0·8 (47)
Number seeds eaten	15	59	74 (29)
Seconds/seed[b]	29·2±2·8	24·0±7·7	28·1±2·7
Number seeds dropped	3	7	10

[a] 1975 – 24 untreated and 12 rows of seed treated with methiocarb. 1976 – 18 untreated rows and 18 treated rows.
[b] Means±S.E.

Though Mesurol^R is registered in the United States to protect sprouting corn and ripening tart and sweet cherries from bird damage, the most definitive test of its repellent properties in the field was only recently conducted and the data from this test are just now becoming available. Mott *et al.* (1979 unpublished) conducted a large-scale field test of the efficacy of Mesurol^R at protecting wine grapes from bird depredations. During September and October 1978, they treated one-half of each of eight commercial vineyards with Mesurol^R and left the other half as an untreated control. Visual estimates were made of the bird damage to each of five bunches on 50 vines in each half of each vineyard. These resulted in a damage rating of 0·72 in the treated units and 2·57 in the untreated units.

Conclusion

The extensive literature on rats, experiments with birds reported here and the literature on the chemical defences of plants to herbivores, reviewed in this symposium and elsewhere (Rogers 1978a), lead to the inescapable conclusion that the conditioned aversion mode of action will be emphasised in developing effective taste-acting repellents to protect agricultural crops from birds. About the use and development of such repellents we can say the following:

1. A low level of damage must be expected during the conditioning period of the pest population; 100% protection is not a reasonable expectation.

2. Because of differences between crops, a repellent that is effective at protecting one will not necessarily protect others; physical characteristics of crops may dictate differing rates or techniques of consumption e.g., hulling of topically treated seeds by the depredating species.
3. It is very unlikely that any one repellent will be effective against all pests because of differences in behaviour and physiology of the species involved. It is unreasonable to expect a panacea.
4. All the presently viable vertebrate repellents involve some adverse effects on the physiology of the target species as an important part of their mode of action; it is likely that future repellents will possess similar attributes.
5. Effective repellents are not necessarily bad-tasting; the pest learns to associate the particular taste with an adverse physiological effect.
6. The use of repellents demands a behavioural response from the target animals, the response being an alteration in feeding behaviour.
7. Any repellent is likely to be most effective when adequate alternative foods are available. Conditioned aversion is the mode of action most likely to produce the required alteration of food habits.

References

ALCOCK, J. (1970) Punishment levels and the response of black-capped chickadees to three kinds of artificial seeds. *Animal Behaviour* 18, 592–599.

ARMOUR, C. J. (1963) Use of repellents for preventing mammal and bird damage to trees and seed. *Forestry Abstracts* 24, XXXVII–XXXVIII.

BESSER, J. F.; WELCH, J. F. (1959) Pesticides and controls; Chemical repellents for the control of mammal damage to plants. *Transactions of the Twenty-fourth North American Wildlife Conference* 24, 166–173.

ROGERS, J. G., JR. (1974) Responses of caged red-winged blackbirds to two types of repellents. *Journal of Wildlife Management* 38, 418–423.

ROGERS, J. G. JR., (1978a) Repellents to protect crops from vertebrate pests: Some considerations for their use and development. In: Bullard, R. W. (*Ed*) Flavor Chemistry of Animal Foods. (American Chemical Society Symposium Series) 67, 150–165.

ROGERS, J. G., JR. (1978b) Some characteristics of conditioned aversion in red-winged blackbirds. *The Auk* 95, 362–369.

ROGERS, J. G. JR; LINEHAN, J. T. (1977) Some aspects of grackle feeding behaviour in newly planted corn. *Journal of Wildlife Management* 41, 444–447.

ROZIN, P.; KALAT, J. W. (1971) Specific hungers and poison avoidance as adaptive specializations of learning. *Psychological Reviews* 78, 459–486.

RZOSKA, J. (1953) Bait shyness, A study in rat behaviour. *The Journal of Animal Behaviour* 1, 128–135.

WELCH, J. F. (1954) Review of chemical repellents for rodents. *Journal of Agricultural and Food Chemistry* 2, 142–149.

WILCOXON, H. C.; DRAGOIN, W. B.; KRAL, P. A. (1971) Illness-induced aversions in rat and quail: relative salience of visual and gustatory cues. *Science* 171, 826–828.

Food selection by skylarks: the effect of a pesticide on grazing preferences

Rhys Green*
Broom's Barn Experimental Station
Bury St Edmunds

Summary

Skylarks foraging in sugar beet fields in spring graze the crop seedling cotyledons causing economic damage in some cases. The birds also take weed seeds and beetles from the soil surface and graze weed seedlings. Counts of grazed seedlings, and micro-analysis of bird faeces, both showed that skylarks preferred sugar beet seedlings to the available weed seedlings. However, where crops were treated with aldicarb, a seed furrow granular pesticide, this preference was less apparent. In cage studies, captive skylarks grazed fewer aldicarb-treated seedlings than untreated seedlings. A survey of commercial plantings of sugar beet also showed significantly less grazing on aldicarb-treated than untreated crops.

Introduction

In eastern England the grazing of sugar beet seedlings by birds has occurred for many years but has caused greater concern to farmers in the 1970s than hitherto. The incidence of grazing may have increased but changes in crop husbandry have also altered the conspicuousness and potential economic impact of damage to the crop in the early stages of growth (Hull & Jaggard 1971). The damage occurs mainly to seedlings in April and May and consists of the loss of all or part of the cotyledons. The growing point is not usually damaged and the plants survive but the rate of early growth is reduced and considerable reductions in yield can occur (Green 1978a).

Several bird species graze sugar beet seedlings but the most important is the skylark (*Alauda arvensis*) (Dunning 1974, Dunning & Green 1975, Green 1978a). Skylarks are common breeding birds on farmland where they nest on the ground amongst the crops. They are particularly abundant in arable farming areas where cereal growing predominates (Williamson 1967, Green 1978a). On arable farmland their food includes grain, weed seeds, arthropods (particularly beetles) and the leaves and cotyledons of weed and crop seedlings (Hammond 1912, Green 1978b).

In April and May, when damage to sugar beet seedlings occurs, the skylark population to be found on any farm will consist of locally breeding adults. The birds are not exclusively territorial and tend to forage where food is most plentiful (Green 1978a,b). A detailed study of the factors affecting the abundance and diet of skylarks on sugar beet fields will be published elsewhere, but the results of this study are summarised here as a background to the subject of the present paper.

Densities of foraging skylarks on fields of seedling sugar beet are very variable and are positively correlated with the density of palatable seedlings of both weeds and crop; these form the bulk of the food on most fields. Beetles and weed seeds, taken from the soil surface, can also be important foods where they are abundant. They are preferred to

*The author is currently working at the Game Conservancy, Fordingbridge, Hampshire.

seedling cotyledons and tend to replace these in the diet as their density increases. In particular, on fields with high densities of weed seeds in the soil, skylarks graze a smaller proportion of the seedlings than on fields where seeds are scarce. As a result, the rate at which individual skylarks graze sugar beet seedlings declines as the density of weed seeds increases. However, since the densities of seeds and seedlings are strongly intercorrelated, and since birds aggregate where seedlings are most plentiful, the effect of seed and seedling density on the total grazing rate is complex. Damage to the crop may be least severe where weed seeds and palatable seedlings are *either* very rare or very common. In the first case each bird is grazing at a high rate but bird density is low since the total food supply on the field is small. In the second case the density of birds is high but they are feeding mainly on seeds rather than seedlings. The most severe damage to the crop might be expected at intermediate densities of these foods.

Cotyledons of weed seedlings are at least as important in the diet as those of sugar beet. The different weed species are not grazed in proportion to their abundance and the preferences which exist are consistent from field to field. Preference rankings are related to the nitrogen and phosphorus content of the seedlings; those species with the highest concentrations being preferred. Hence, skylarks may, by grazing selectively, enhance their intake of certain nutrients, perhaps protein. Similar selective feeding is recorded in other herbivorous birds (Moss 1972, Mills & Mark 1977, Summers & Jones 1976). The possibility that the presence of aversive secondary substances in certain seedling species influenced grazing preference was not investigated but there was considerable variation which was not accounted for by the concentrations of measured nutrients. Furthermore, studies of insect food selection demonstrate the general importance of these substances (Bernays & Chapman 1977).

Consideration of three factors which might affect the response of skylarks to sugar beet seedlings, (i) the abundance of seeds as preferred, alternative food, (ii) the nutrient content and (iii) the presence of secondary substances, suggests ways in which the birds' tendency to graze sugar beet seedlings might be artificially reduced. Firstly, herbicides might be used to regulate supplies of weed seeds and seedlings and thus affect damage. However, as we have seen, the effects of weed abundance act in opposite directions on the density of skylarks and their predeliction for seedlings so the results of such changes would be difficult to predict. Secondly, the nutritional value of seedlings might be changed by using particular varieties or by modifying fertiliser application. Summers & Jones (1976) found that varieties of pear trees whose buds had the lowest nitrogen content suffered least from the attacks of bullfinches whilst Miller (1968) and Owen (1975) showed that red grouse and geese grazed selectively on plots of their food plants which had been given supplementary fertiliser. Thirdly, the levels of aversive substances in sugar beet seedlings might be increased by selective breeding or by supplying exogenous deterrent chemicals. Some wild plant populations are known to have genetic polymorphisms for chemical deterrents (Daday 1954) and sugar beet might have such materials whose concentration could be increased under selection. Artificial repellents may reduce plant palatability but, if applied as surface treatments, have the disadvantage that they can be washed away. A more satisfactory type of artificial deterrent would be taken up by the plant and either be aversive to grazing animals in itself or stimulate the production of aversive substances by the plant. Carbamates, particularly methiocarb, have been used as surface repellents against birds on a variety of crops (Guarino 1972, Porter 1977). In this paper I present evidence that a dithiocarbamate pesticide, acting systemically on sugar beet seedlings, reduces their palatability to skylarks. The results are divided into three sections, the first deals with the grazing preferences of wild skylarks foraging in sugar beet fields, the second with feeding trials with captive skylarks, and the third with a survey of levels of grazing damage occurring to commercial crops of sugar beet.

1. The effect of aldicarb on the palatability of sugar beet seedlings to wild skylarks

Methods

Nine sugar beet fields sown in late March – early April 1978 in Cambridgeshire and West Suffolk were studied. Four of the fields were treated at sowing with a granular formulation of aldicarb (Temik 10G) in the seed furrow at rates within the manufacturer's recommended range (0·5–1·0 kg a.i. ha^{-1}). The aldicarb, buried in the seed furrow, protects the germinating seed from soil invertebrates; it is also absorbed by the young seedlings and kills aphids which feed on the plant for several weeks after emergence. The other five fields received no seed furrow treatment. All the fields, which were on five widely separated farms, received a variety of herbicide and fertiliser treatments.

Birds were counted and skylark faeces collected regularly throughout the cotyledon stage of the beet seedlings. Mean densities were calculated for each bird species for each field and the diet of skylarks determined by micro-analysis of faecal pellets. The species composition of the seedling cotyledons eaten was estimated by identifying species from the cell morphology of epidermal fragments in the faeces and measuring the contribution made by each to the total area of seedling epidermal fragments. Full details of these methods can be found elsewhere (Green 1978b).

At the end of the cotyledon stage, 13–21 days after emergence, densities of crop and weed seedlings, and the levels of grazing on them, were estimated. Sugar beet seedlings were counted in 30 lengths of row, each of 10 m and weed seedlings in 30 quadrats measuring 0·1m^2. Sampling was done by firstly pacing out a regular grid of 30 points on the field; from each point a stick was then thrown at random to determine the precise position for the count. Each seedling was scored according to the severity of grazing as follows;

0	no damage
1 ~	25% of cotyledon area missing
2 ~	50% of cotyledon area missing
3 ~	75% of cotyledon area missing
4 ~	100% of cotyledon area missing

Red-legged partridges (*Alectoris rufa*) and woodpigeons (*Columba palumbus*) were the only grazing birds, other than skylarks, present on sugar beet fields and their combined densities were low (mean 0·09: range 0–0·47 birds ha^{-1}) compared with the density of skylarks (mean 1·06: range 0·52–3·42 birds ha^{-1}). Therefore, it is assumed that most of the observed grazing was due to skylarks. Furthermore, enclosure on these, and other fields, showed that grazing by animals other than birds was relatively unimportant (Dunning & Green 1975).

Results

Seedlings of four weed species commonly grazed by skylarks were present on the fields. The relative preference of skylarks for sugar beet seedlings compared with these palatable weed seedlings was calculated by two methods.

(a) *Faecal analysis*

The mean area of cotyledon epidermis of each seedling species, as a proportion of the total area of epidermis of all the species under consideration, f, was obtained for each field. The relative risk of sugar beet seedlings being grazed, compared with the palatable weed seedlings, was calculated as:

$$RR_1 = \frac{\frac{f_1}{d_1}}{\frac{f_1}{d_1} + \frac{f_2}{d_2} + \frac{f_3}{d_3} + \frac{f_4}{d_4}}$$

where RR is relative risk and d the density by dry weight of a particular seedling species. The subscripts refer to: (1) sugar beet (2) *Polygonum aviculare* (3) *Polygonum convolvulus* (4) *Atriplex patula* and *Chenopodium album*. The results for the last two species had to be combined because the remains of epidermis in the birds' faeces could not be separately identified.

(b) *Counts of grazing intensity*

The relative risk of grazing for sugar beet seedlings was calculated as:

$$RR_1 = \frac{p_1}{p_1 + p_2 + p_3 + p_4}$$

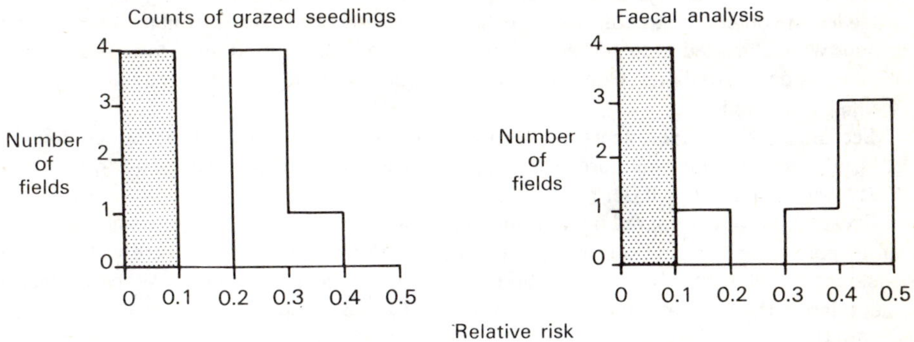

Figure 1. *Relative risk of grazing by skylarks for sugar beet seedlings relative to palatable weed seedlings*. Numbers of aldicarb-treated (stippled) and untreated fields with different values of relative grazing risk for sugar beet seedlings (see text), assessed by (A) counts of grazed seedlings (B) analysis of skylark faeces.

where for each seedling species p is the proportion of the total cotyledon area removed. Subscripts refer to species as before. These formulae for relative risk follow those used by Goss-Custard (1977 Appendix I).

Relative risk values calculated for each field are shown in Fig. 1. The relative risk of grazing for sugar beet seedlings, compared with palatable weed seedlings differed significantly (Mann Whitney U test $P<0{\cdot}05$) between aldicarb-treated and untreated fields, being lower on the treated fields. The preference ranking among the weed species was not affected by aldicarb treatment. I conclude that the palatability of sugar beet seedlings was reduced on the aldicarb-treated fields.

Because of this difference in palatability the proportion of the cotyledon area removed by grazing was significantly less on aldicarb-treated than untreated fields (mean 2·4% range 0·2–4·0% cf. mean 25·0% range 12·9–51·3% Mann Whitney U test $P\leqslant0{\cdot}05$). Mean skylark densities on treated and untreated fields were not significantly different.

2. The effect of aldicarb on the palatability of sugar beet seedlings to captive skylarks

Methods

Six male skylarks were trapped and confined in groups of three in two outdoor aviaries. Birds were fed on turkey pellets, with a vitamin and mineral supplement, and had access to *ad lib.* supplies of this food and to water throughout the study. Fifteen days after capture each group was presented with three trays of newly-emerged sugar beet seedlings; one group receiving seedlings treated with aldicarb, the other, untreated seedlings. Ten days later fresh trays of seedlings were presented, the treatments being reversed. This exposure of the birds to treated and untreated seedlings was intended to give them experience of both types prior to the commencement of the main experiment.

The birds were then caged individually in outdoor flights and given a single tray of seedlings for five days followed by a fresh tray for a further five days. Three birds received untreated seedlings for the first five-day period and aldicarb-treated plants for the second, while for the others the treatments were given in reverse order. Birds from the two original groups were allocated to the different experimental regimes at random. At the end of each of the five-day periods the trays were removed and the numbers of grazed and ungrazed seedlings counted.

Seedlings for the experiments were grown in $40\times21\times5$ cm polystyrene trays containing 2 kg of a soil-compost mixture consisting of sterilised loam, peat and sand in the ratio 7:3:2 with John Innes base.

Treatments were prepared by thoroughly mixing Temik 10G with air-dry soil at the rate of 10 ppm a.i. before the soil was placed in trays. After sowing, trays were watered and enclosed in polythene; being left to germinate without further attention. When placed in the aviaries the trays contained an average of 61 seedlings.

Results

Table 1 shows the proportions of treated and untreated seedlings grazed by individual skylarks. In all cases the birds grazed a larger proportion of untreated than treated seedlings (Wilcoxon matched pairs test $p<0{\cdot}05$). I conclude that captive skylarks, with

Table 1

Proportion of aldicarb-treated and untreated seedlings grazed in five days by each of six captive skylarks.

Order of presentation of treatments	% Seedlings grazed	
	Untreated	Aldicarb-treated
Untreated first	38·2	1·6
Untreated first	92·1	1·7
Untreated first	20·7	0
Aldicarb first	19·6	11·8
Aldicarb first	21·3	7·4
Aldicarb first	7·5	2·7

continuous access to alternative food, are less likely to eat aldicarb-treated sugar beet seedlings than untreated control plants.

3. Grazing damage to aldicarb-treated and untreated sugar beet crops

Methods

Observations were made in 1976 on sugar beet fields situated throughout eastern England and the West Midlands. After drilling, but before germination, 10 sample lengths of crop row, each measuring 10 m, were selected at random and marked with canes. The fields were revisited at the end of the cotyledon stage when the seedlings in the sample lengths were counted and scored for grazing damage by the method given in section I. The proportion of the total cotyledon area removed by grazing was calculated for each field.

Results

Table 2 shows the numbers of fields with different levels of grazing damage for aldicarb-treated fields and those with no seed furrow pesticide treatment. The fields received a wide variety of herbicide and fertiliser treatments and were on a range of soil types. The 17 fields treated with seed furrow aldicarb (Temik 10G at 0·5–1·0 kg a.i. ha^{-1}) suffered significantly less grazing damage than untreated fields (Fisher exact probability test $p = 0{\cdot}004$). However, within treated fields there was no correlation between damage intensity and the rate of application of the pesticide.

Table 2

Grazing of sugar beet cotyledons in relation to seed furrow aldicarb treatments

	Untreated	Aldicarb
number of fields with:		
0–5% cotyledon area grazed	33	16
6–10% cotyledon area grazed	9	1
> 10% cotyledon area grazed	17	0
mean % area grazed	7·8	2·1
range % area grazed	0–44	0–6

Discussion

Grazing pressure by skylarks on sugar beet seedlings, compared with that on palatable weed seedlings, was reduced on fields treated with aldicarb in the seed furrow and this was apparent both when the effect was measured directly by scoring the damage, and by analysis of the species composition of seedlings grazed. Since the order of preference for seedlings, other than those of sugar beet, was relatively constant from field to field and unaffected by aldicarb treatment, this result indicates that the treatment reduced the palatability of the crop but not the weed seedlings. The specificity of this effect might be due to species differences in the uptake or metabolism of aldicarb or, more probably, to the fact that it is only applied close to the beet seedlings. Most weed seedlings grew between the crop rows, which were 50 cm apart, and therefore weeds were probably exposed to a much lower concentration of aldicarb around their roots than the sugar beet.

The reduction in grazing damage on the aldicarb-treated fields was not due to the birds moving to feed on other fields or being killed by the pesticide since there was no significant reduction in skylark density. A reduced rate of grazing on sugar beet by each bird, rather than lower bird numbers, seems to explain the results best. Lethal or sub-lethal effects of aldicarb on skylarks were not examined and are not excluded by this study; however, such effects are not required in order to explain the reduction in crop damage observed (P. J. Bunyan pers. comm.).

The survey of grazing damage on a large sample of fields confirmed that aldicarb treated crops suffered less damage than those not treated with a seed furrow pesticide. It might be that fields selected by growers for aldicarb treatment differed from untreated fields in some other way which resulted in different damage levels. This argument seems unlikely to be valid since aldicarb was in widespread use in a large proportion of fields in 1976. The treated fields in the survey were not aggregated in a particular geographical area or on particular soil types. Similarly, the treated and untreated fields used for the study of grazing preferences were fairly well matched for soil type, weed flora and general environment.

The feeding trials with captive skylarks confirm that these birds tend to avoid feeding on aldicarb-treated sugar beet seedlings. The treated and untreated seedlings were not obviously different in appearance and it seems likely that the birds responded to the taste of the cotyledons. The concentration of aldicarb used in these trials was chosen to be similar to that around the roots of seedlings under field conditions but the extent of diffusion of aldicarb in soil has not been sufficiently studied to know whether the correspondence was close.

The tendency of skylarks to graze sugar beet seedlings, in addition to being affected by the availability of preferred foods (Green 1978a), is altered by treatment with the dithiocarbamate pesticide aldicarb. Since this pesticide is buried in the soil, any effect on the palatability of the seedling cotyledons must be by a systemic action. Given the sporadic and unpredictable occurrence of economically significant damage by grazing birds, it is unlikely that prophylactic application of present formulations of aldicarb to sugar beet would repay the high cost of treatment. However, the mode of action, particularly the possibility of systemic effects, of this, and other carbamate compounds, in repelling birds, seems worthy of further investigation.

Acknowledgements

I am grateful to R. A. Dunning and E. N. Wright for help and advice, the fieldstaff of the British Sugar Corporation for help in the field and Gillian Davison for typing. The work was supported by the Ministry of Agriculture's Sugar Beet Research and Education Committee.

References

BERNAYS, E. A.; CHAPMAN, R. F. (1977) Deterrent chemicals as a basis of oligophagy in *Locusta migratoria* (L.), *Ecological Entomology,* 2, 1–18.

DADAY, H. (1954) Gene frequencies in wild populations of *Trifolium repens. Heredity,* 8, 61–78, 377–384.

DUNNING, R. A. (1974) Bird damage to sugar beet. *Annals of Applied Biology,* 76, 325–366.

DUNNING, R. A.; GREEN, R. E. (1975) Bird damage to sugar beet and approaches to its control. *Proceedings 8th British Insecticide and Fungicide Conference 1975,* 1, 237–244.

GOSS-CUSTARD, J. D. (1977) The energetics of prey selection by redshank *Tringa totanus* (L) in relation to prey density. *Journal of Animal Ecology,* 46, 1–19.

GREEN, R. E. (1978a) The ecology of bird and mammal pests of sugar beet. Ph.D. Thesis, University of Cambridge.

GREEN, R. E. (1978b) Factors affecting the diet of farmland skylarks *Alauda arvensis. Journal of Animal Ecology,* 47, 913–928.

GUARINO, J. L. (1972) Methiocarb, a chemical bird repellent: a review of its effectiveness on crops. In: Marsh, R. E. (*Ed*) *Proceedings 5th Vertebrate Pest Control Conference,* Davis; University of California. p. 108–111.

HAMMOND, J. (1912) An investigation concerning the food of certain birds. *Journal of Agricultural Science, Cambridge,* 4, 380–409.

HULL, R.; JAGGARD, K. W. (1971) Recent developments in the establishment of sugar beet stands. *Field Crop Abstracts,* 24, 381–390.

MILLER, G. R. (1968) Evidence for selective feeding on fertilised plots by red grouse, hares and rabbits. *Journal of Wildlife Management,* 32, 849–853.

MILLS, J. A.; MARK, A. F. (1977) Food preferences of the takahe in Fiordland National Park, New Zealand, and the effect of competition from introduced red deer. *Journal of Animal Ecology,* 46, 939–958.

MOSS, R. (1972) Food selection by red grouse (*Lagopus lagopus scoticus* (Lath.)) in relation to chemical composition. *Journal of Animal Ecology,* 41, 411–428.

OWEN, M. (1975) Cutting and fertilizing grassland for winter goose management. *Journal of Wildlife Management,* 39, 163–167.

PORTER, R. E. R. (1977) Methiocarb protects sprouting peas from small birds. *New Zealand Journal of Experimental Agriculture,* 5, 335–338.

SUMMERS, D. D. B.; JONES, F. J. S. (1976) The importance of protein in the selection of fruit buds by bullfinches. *Experimental Horticulture,* 28, 47–50.

WILLIAMSON, K. (1967) The bird community of farmland. *Bird Study,* 14, 210–226.

Plant secondary compounds as a chemical defence

L. E. Fellows
Jodrell Laboratory
Royal Botanic Gardens, Kew

Summary

Chemicals produced by plants are now believed to play a vital rôle in their interaction with other organisms. This interaction is of interest because plant chemicals (a) include a wide range of biodegradable pesticides (e.g. fungicides, insecticides) which may be exploited in agricultural practice, and (b) largely determine the attractiveness or otherwise of a potential food plant to a predator.

Primary and secondary plant chemicals

The biochemical reactions which support life are fundamentally the same in all living organisms, and the chemical compounds which participate in those reactions are universally distributed and known as PRIMARY COMPOUNDS. Examples of primary compounds are simple sugars, protein amino acids, simple fatty acids and the common purine and pyrimidine bases. Also included in the definition of primary compounds are those chemicals which participate in some vital, specialised function, such as photosynthesis in plants and cell wall synthesis in bacteria. SECONDARY COMPOUNDS, (SCs), in contrast, are defined as those having no obvious rôle to play in life-supporting biochemistry. The greatest variety of types of SC are found in plants and micro-organisms, where a particular compound is usually of restricted distribution, i.e. it is confined to a closely related group of species and is absent from others. Modern techniques in chemistry have simplified the extraction and identification of novel SCs and hundreds of newly discovered structures are reported each year. It is less easy to explain why they occur at all, and in such amazing diversity. The competition between plants for light and nutrients is such that it is inconceivable that any species which squandered its resources in synthesising compounds for no purpose would survive the evolutionary rat-race. A possible explanation is that SCs confer on the plant some strategic advantage, e.g., by acting as a deterrent to pathogenic micro-organisms, phytophagous insects, grazing mammals and competing plant species, or alternatively by attracting desirable organisms, such as pollinating insects.

A wide range of SC types have now been isolated and demonstrated to play a rôle in plant chemical ecology, e.g., alkaloids, flavonoids, phenolics, terpenes, non-protein amino acids, polyacetylenes, glucosinolates, lectins, etc. However, it is not possible to predict from the structure of any particular SC exactly what its rôle will be. Many different kinds of SC may have similar effects, e.g. certain tannins, alkaloids, lectins, flavonoids, terpenoids and non-protein amino acids have been shown to be insect deterrents (Harborne 1977). A particular compound may, at the same time, be repellent to one organism and attractive to another. The mustard oil, allyl isothiocyanate, released from cabbage is fungitoxic but attracts the cabbage root fly *Delia brassicae* to its host plant (Finch 1978). A compound may also be deterrent to several different kinds of predator.

The non-protein amino acid canavanine, which occurs in some legumes, is toxic to insects, mammals and to other plants (Bell 1978). It is observed that most plants only accumulate one or two SCs at high concentration, which probably fulfil most of the requirements of chemical strategy. Presumably no species can afford the metabolic cost of producing a variety of defensive compounds.

Chemical "struggle" in evolution

The rise of the vascular plants began 400 million years ago and during their early evolution micro-organisms and insects were the main predators. Mammals, and particularly man, are a relatively recent menace (see Table 1). For most classes of plant SC, increasing structural complexity parallels evolutionary advancement, e.g. the flavonoids increase in both number and type from mosses and liverworts through ferns and gymnosperms to angiosperms (Swain 1975). This is now believed to be the result of a continuous struggle between plants and other organisms. As a chance mutation in a plant enabled it to synthesise a more effective deterrent to, e.g. an insect predator, so that species increased in number relative to other species with less effective deterrents. The result of this success, however, was an increased likelihood that some predator would acquire resistance to that deterrent. Expansion of the plant group was thus checked until further chemical "inventiveness" allowed it to expand again (Cronquist 1977). Thus, as evolution progressed, there was selection for increasing complexity and variety in plant SCs, i.e. even more effective antibiotics, insecticides, herbicides, pollinator attractants, etc.

Table 1

Summary of Plant-Animal Co-evolution

Million years BP	Plants	Animals
0·01	Cultivated plants	
2·5		Origin of man.
135	Woody angiosperms, grasses and pines.	Rise of mammals. End of giant reptiles.
180	Origin of angiosperms. End of seed ferns.	Birds and higher insects.
325	Seed ferns, Gymnosperms.	Reptiles, insects.
400	Rise of vascular plants, inc. ferns.	Fish.
600	Marine algae.	Marine invertebrates.
3500	Bacteria, algae.	
4500	Age of oldest rocks.	

(Adapted from Swain 1974)

Avoidance of self-poisoning

Any SC produced by a plant for defence must be non-toxic to the species in which it accumulates. Several solutions to this problem are evident. A common device is to store the toxin in a harmless form, often as a glycoside. The resistance of young *Sorghum bicolor* to attack by the locust *Locusta migratoria* is partly due to the HCN produced when the leaves are bitten (Fig. 1). The cyanogenic glycoside dhurrin, harmless in itself, then comes into contact with an enzyme normally stored in a different part of the cell, and cyanide is released (Woodhead & Bernays 1977).

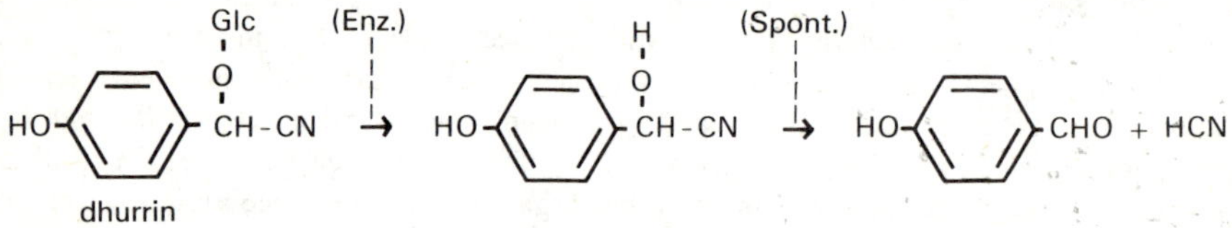

Figure 1. Release of HCN from dhurrin in *Sorghum bicolor*

Similarly, (Fig. 2) the phytotoxin juglone accumulates in the leaves of the black walnut, *Juglans nigra* as a harmless glycoside. This is leached from the leaves and breaks down in the soil where its product inhibits the growth of many plant species (Bode 1958).

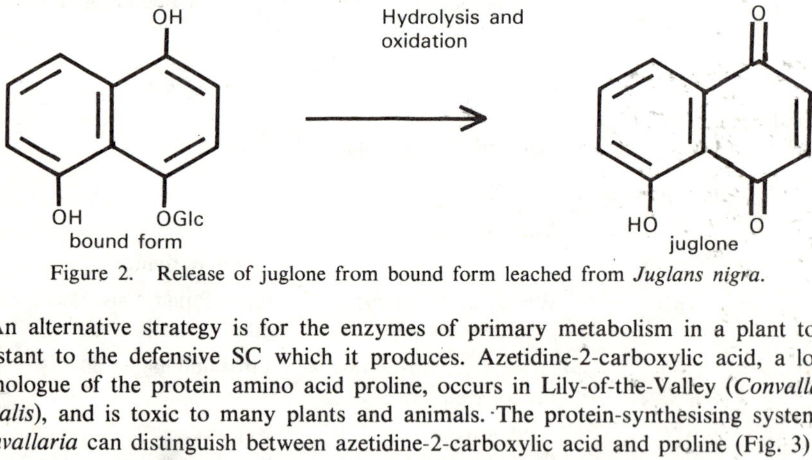

Figure 2. Release of juglone from bound form leached from *Juglans nigra*.

An alternative strategy is for the enzymes of primary metabolism in a plant to be resistant to the defensive SC which it produces. Azetidine-2-carboxylic acid, a lower homologue of the protein amino acid proline, occurs in Lily-of-the-Valley (*Convallaria majalis*), and is toxic to many plants and animals. The protein-synthesising system of *Convallaria* can distinguish between azetidine-2-carboxylic acid and proline (Fig. 3) but those of susceptible species cannot and, as a result, produce defective protein (Fowden 1963).

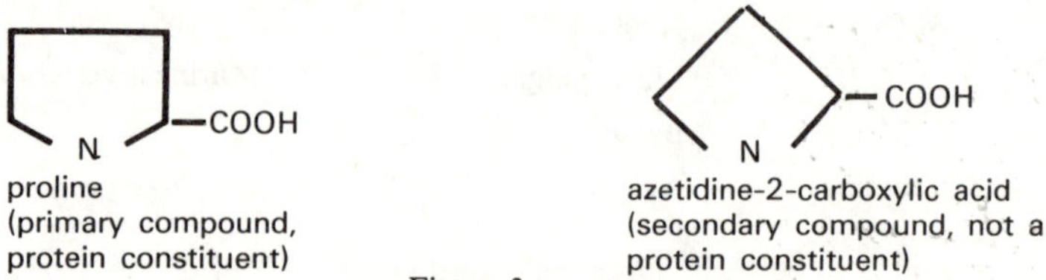

Figure 3

Variety in chemical strategy

It is in the interest of the plant to produce no more of any SC than is necessary. Many plant SCs, particularly phenolics, confer resistance to fungal infection, but a particularly interesting (and heterogeneous) group of compounds known as phytoalexins are only formed when the plant is directly under attack. Thus these are only found in living cells, and are confined to the tissue colonised by the fungus and its immediate neighbourhood. Each plant family produces its own characteristic type of phytoalexin molecule (Fig. 4), usually structurally related to other SCs characteristic of that family (Harborne & Ingham 1978).

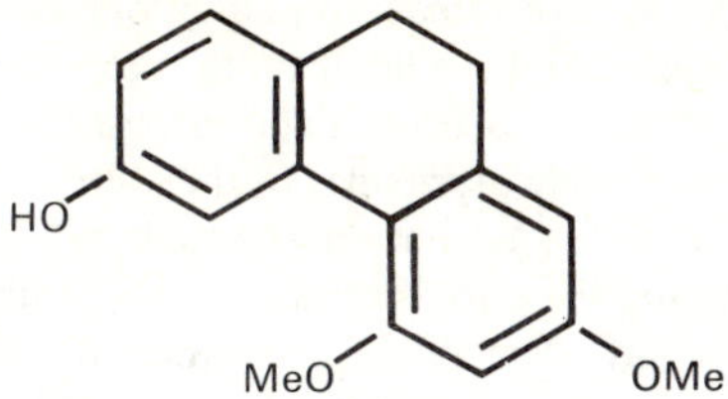

orchinol
(*Orchis militaris*)

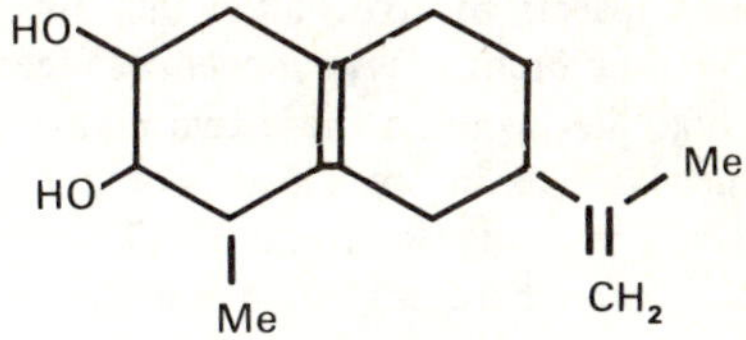

rishitin
(*Solanum tuberosum*)

$HOCH_2CHOHCH{=}CH(C{\equiv}C)_3CH{=}CHMe$
safynol
(*Carthamus tinctoria*)

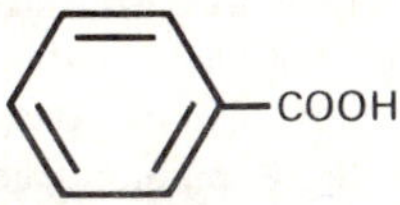

benzoic acid
(*Malus pumila*)

Figure 4. Some phytoalexins of higher plants.

The metamorphosis of insects from larva to adult is controlled by hormones. Some plants, mainly ferns and gymnosperms, synthesise massive amounts of these hormones which interfere with the development of insect predators. Originally, half-a-ton of silkworms was processed to yield 25 mg of insect moulting hormone: this proved to be a steroid, α-ecdysone, together with lesser amounts of the closely related β-ecdysone (Karlson *et al.* 1965). Chemists were then amazed to find high levels of β-ecdysone in the Yew, *Taxus baccata,* where only 25 g of dried leaf was sufficient to yield 25 mg β-ecdysone, and in the common fern *Polypodium vulgare,* where 2·5 g of rhizomes sufficed. Insects can, however, degrade these natural hormones. Many plants have now been found to synthesise structural analogues (see Fig. 5) which are less easily degraded

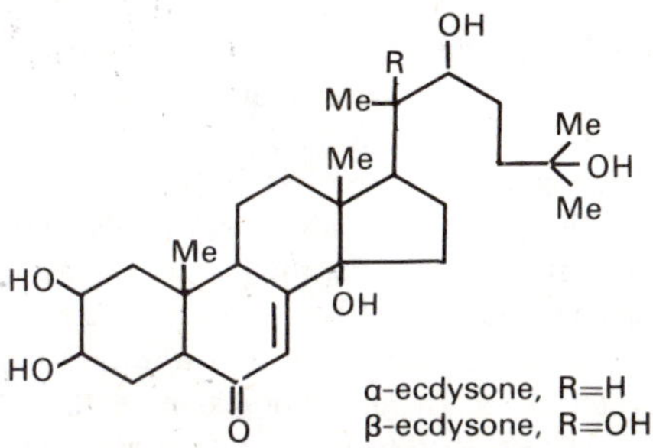

α-ecdysone, R=H
β-ecdysone, R=OH

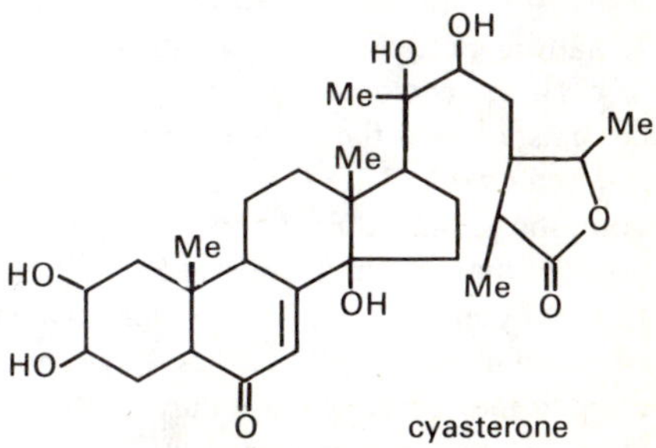

cyasterone

Figure 5. Insect moulting hormones and a phytoecdysone from *Cycas*.

and therefore effective at far lower concentration. Over 30 such phytoecdysones have been reported (Harborne 1977). This brilliant strategy for dealing with insect pests was "invented" by plants tens of millions of years ago (Williams 1972).

Complex interactions

Since plants have been "experimenting" with SCs for millions of years, it is not surprising that the chemical interactions between organisms are often complex. The successful weeds groundsel (*Senecio vulgaris*) and ragwort (*S. jacobaea*) are protected from mammals and most insects by toxic alkaloids, but caterpillars of the tiger moth (*Arctia caja*) and the cinnabar moth (*Tyria jacobaeae*) feed with impunity on both weeds and carry out their whole life cycle on these two plants. Both the caterpillars and the adult moths sequester the alkaloids in specialised compartments: these then serve to protect the insects from attack by birds (Rothschild 1972). Similarly, the toxic cardiac glycosides of the milkweed (*Asclepias curassavica*) are sequestered by the caterpillars of butterflies, such as the monarch (*Danaus plexippus*), which live on the plant and these toxins persist in the adult, which is then protected from predation by the blue jay. The bright coloration of the butterfly serves as a warning to the birds, which then avoid similarly coloured butterflies whether or not they contain toxins (Roeske *et al.* 1976).

Most wild legumes contain SCs which are toxic to herbivores, e.g. seeds of many species of the Papilionoideae contain canavanine (Fig. 6), a non-protein amino acid toxic to a wide range of organisms, including most insects. Seeds of *Dioclea megacarpa* are the exclusive food of the bruchid beetle *Caryedes brasiliensis,* although they contain 8% canavanine. This bruchid has adapted its metabolism to avoid the toxic effects of canavanine, which it is alleged to break down and use as food (Rosenthal *et al.* 1976). A recent study of Central American legumes has revealed that most species are the exclusive hosts of one or two bruchid species, each bruchid having overcome the defences of its host, but no others (Janzen 1978).

$_2HN\text{-}C(NH)\text{-}NH\text{-}CH_2\text{-}CH_2CH_2CH(NH_2)COOH$ arginine, (a protein amino acid)

$_2HN\text{-}C(NH)\text{-}NH\text{-}O\text{-}CH_2CH_2CH(NH_2)COOH$ canavanine (a non-protein amino acid)

Figure 6. Canavanine, a structural analogue of arginine.

The anomalous position of crops

In natural circumstances populations of living organisms remain fairly constant as a result of balanced ecological pressures, including chemical pressure. One instance of unnatural circumstances is the growth by man of crops on a large scale. Cultivated plants have been selected over c. 10 000 years for high yield and low toxicity to mammals, i.e. man and his domestic animals. Thus those strains were selected in which biosynthetic resources were diverted away from protective SCs into primary storage compounds, such as starch. As a result of continuous selection, our main food crops have lost the chemical defences still to be found in their wild relatives. For example, seeds of the common vetch (*Vicia sativa*) are toxic to mammals because they contain a non-protein amino acid, β-cyanoalanine; this is absent from the broad bean (*V. faba*) (Bell 1972). The risk to crops is increased by the practice of large-scale mono-culture, which encourages the proliferation of predators. This

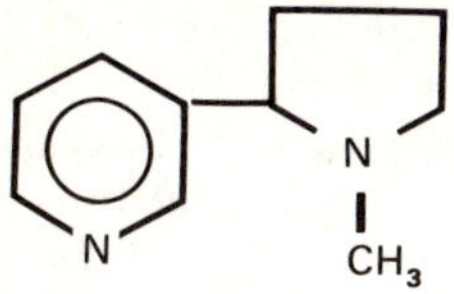

Nicotine
(ex *Nicotiana tabacum*)

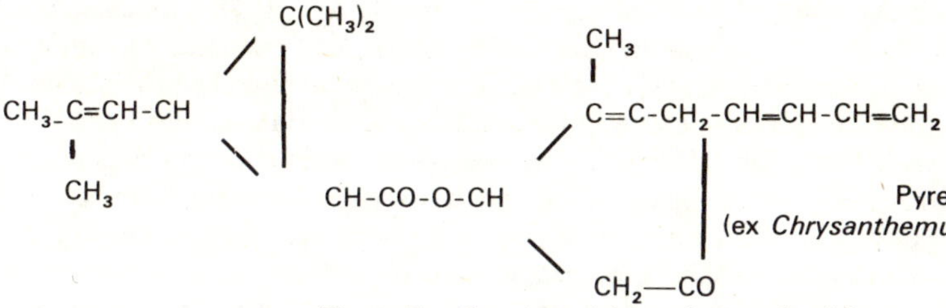

Pyrethrin I
(ex *Chrysanthemum cinerariaefolium*)

Figure 7. Two insecticides of natural origin.

has resulted in the present feverish attempts by human chemists to devise protective chemicals to replace those lost. Apparent success has sometimes been marred by unwanted side-effects: e.g. dieldrin proved toxic to a far wider range of organisms than those for which it was intended, and, being non-biodegradable, accumulated in soil causing further ecological disturbance.

The future

Plant defensive compounds, which are the result of millions of years "development", are biodegradable and may be limited and specific in their effects. They are thus a potential source of pesticides which might be exploited by man to protect cultivated species without the undesirable side-effects of synthetic chemicals. This is feasible where the SC accumulates at high concentration, e.g. a West African legume (*Griffonia simplicifolia*) contains 14% dry wt of the non-protein amino acid 5-hydroxytryptophan in its seeds (Fellows & Bell 1970). Examples of plant SCs already successfully exploited commercially as insecticides include rotenone (from *Derris* spp.), pyrethrins (from *Chrysanthemum* spp.) and nicotine (from *Nicotiana* spp.) (Fig. 7). Rotenone and pyrethrin are harmless to mammals (Hartley & West 1969). The exploitation of plant SCs by man for agricultural purposes is not new. Certain *Tagetes* spp. release a sulphur-containing thiophene (Fig. 8) into the soil which kills nematodes up to a radius of 3 ft. *Tagetes* flowers have been identified on vase paintings from pre-Inca people of South America who held the plants sacred to the god of agriculture, since they enabled them to grow such crops as potatoes, maize and beans continuously on the same ground for centuries (Wynne Hatfield 1969).

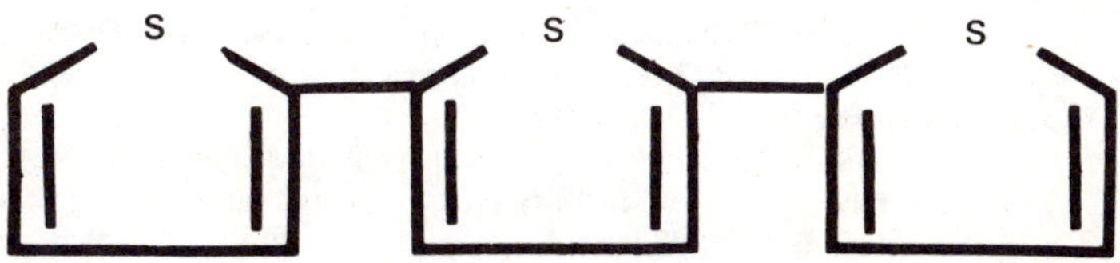

Figure 8. Terthienyl, an anti-nematode compound released from the roots of some *Tagetes* spp.

Although crops have been selected to be low in anti-mammal SCs at harvest, many other SCs remain. Both primary and secondary compounds are known to influence the selection of crop plants as food by predators, both as phagostimulants and deterrents. The acceptability of *Brassica* spp. to the cabbate root fly (*Delia brassicae*) depends on the level of volatile isothiocyanates released (Finch 1978). The Colorado beetle (*Leptinotarsa decemlineata*) responds to the "complete" odour of fully grown potato plants, i.e. all volatiles must be present and in the correct proportions; none of the individual component volatiles alone is attractive (Visser & Avé 1978). Biochemical factors are certainly involved in the selection of food by birds. A better understanding of this aspect of bird predation, largely neglected since ornithologists and chemists seldom collaborate, could prove particularly useful in the selection of resistant strains or in the development of decoy food to lure birds away from crops during periods of maximum vulnerability. Although this approach is unlikely to yield overnight any miracle answer to the problem of bird pests it is free of the environmental hazards attendant on the use of chemical pesticides. The biochemical basis of bird resistance in *Sorghum* strains has been investigated and claimed to be correlated with tannin content, but this is disputed (Perumal & Subramaniam 1973).

Chemical ecology as a scientific discipline owes its existence to an awareness by biologists that chemical factors are important in ecology and the ability of chemists, using modern analytical techniques, to rapidly isolate and identify compounds of interest. While the intrinsic fascination of the subject guarantees its development, the main stimulus has to be the certainty of future economic benefit.

References

BELL, E. A. (1972) Toxic amino acids in the Leguminosae. In: Harborne, J. B. (*Ed*) Phytochemical Ecology. London; Academic Press. p. 163–178.

BELL, E. A. (1978) Toxins in seeds. In: Harborne, J. B. (*Ed*) Biochemical aspects of plant and animal co-evolution. London; Academic Press. p. 143–161.

BODE, H. R. (1958) Beitrage zür Kenntnis allelopathischer erscheinungen bei einigen Juglandaceen. *Planta* 51, 440–480.

CRONQUIST, A. (1977) On the taxonomic significance of secondary metabolites in angiosperms. In: Kubitzki, K. (*Ed*) Flowering plants. Evolution and classification of higher categories. *Plant Systematics and Evolution*. Supplement 1. p. 179–189.

FELLOWS, L. E.; BELL, E. A. (1970) 5-Hydroxy-L-tryptophan, 5-hydroxytryptamine and L-tryptophan-5-hydroxylase in *Griffonia simplicifolia*. *Phytochemistry* 9, 2389–2396.

FINCH, S. (1978) Volatile plant chemicals and their effect on host plant finding by the cabbage root fly, *Delia brassicae*. *Entomologia Experimentalis et Applicata*, 24, 350–359.

FOWDEN, L. (1963) Amino acid analogues and the growth of seedlings. *Journal of Experimental Botany* 14, 387–398.

HARBORNE, J. B. (1977) Introduction to ecological biochemistry. London; Academic Press. xiv + 243 pp.

HARBORNE, J. B.; INGHAM, J. L. (1978) Biochemical aspects of the co-evolution of higher plants with their fungal parasites. In: Harborne, J. B. (*Ed*) Biochemical aspects of plant and animal co-evolution. London: Academic Press. p. 343–405.

HARTLEY, G. S.; WEST, T. F. (1969) Chemicals for pest control. Oxford; Pergamon Press. p. 26–39.

JANZEN, D. H. (1978) The ecology and evolutionary biology of seed chemistry as relates to seed predation. In: Harborne, J. B. (*Ed*) Biochemical aspects of plant and animal co-evolution. London; Academic Press. p. 163–206.

KARLSON, P.; HOFFMEISTER, H.; HUMMEL, H.; HOCKS, P.; SPITELBAR, G. (1965) Reaktionen des ecdysonmoleküls. *Chemische Berichte* 98, 2394–2402.

PERUMAL, F. S.; SUBRAMANIAM, T. R. (1973) Studies on panicle characters associated with bird resistance in *Sorghum*. *Madras Agricultural Journal* 60, 256–258.

ROESKE, C. M.; SEIBER, J. N.; BROWER, J. P.; MOFFIT, C. M. (1976) Milkweed cardenolides and their comparative processing by monarch butterflies. *Recent Advances in Phytochemistry*, 10, 93–167.

ROSENTHAL, G. A.; DAHLMAN, D. L.; JANZEN, D. H. (1976) A novel means of dealing with L-canavanine, a toxic metabolite. *Science* 192, 256–258.

ROTHSCHILD, M. (1972) Some observations on the relationship between plants, toxic insects and birds. In: Harborne, J. B. (*Ed*) Phytochemical Ecology. London; Academic Press. p. 1–12.

SWAIN, T. (1974) Biochemical evolution in plants. In: Florkin M; Stotz, E. H. (*Eds*) Comprehensive biochemistry. Amsterdam; Elsevier. p. 125–302.

SWAIN, T. (1975) Evolution of flavonoid compounds. In: Harborne, J. B.; Mabry, T. J.; Mabry, H. (*Eds*) The flavonoids. London; Chapman and Hall. p. 1096–1129.

VISSER, H. H.; AVÉ, D. A. (1978) General green leaf volatiles in the olfactory orientation of the Colorado beetle, *Leptinotarsa decemlineata*. *Entomologia Experimentalis et Applicata*, 24, 738–749.

WILLIAMS, C. M. (1972) Hormonal interactions between plants and insects. In: Sondheimer, E.; Simeone, J. B. (*Eds*) Chemical ecology. New York; Academic Press. p. 103–132.

WOODHEAD, S.; BERNAYS, E. A. (1978) The chemical basis of resistance of *Sorghum bicolor* to attack by *Locusta migratoria*. *Entomologia Experimentalis et Applicata* 24, 123–144.

WYNNE HATFIELD, A. (1969) How to enjoy your weeds. London; Frederick Muller. p. 8–11.

Discussion

O'Connor: I should like to ask Dr Fellows about an alternative explanation for that gradient in cyanogenesis that she referred to. If you get a parallel with January temperatures, is there not a possibility that if the plants growing in the North-East had a high content of cyanogenic chemical then the release of HCN, and consequent self-poisoning, might be brought about by frost damage?

Fellows: This is not actually my work so I cannot really comment, but you raise an interesting point and I agree that it is a possibility. Selection by slugs and snails is not 100% and there must be other factors involved.

Higgons: I should like to ask Dr Rogers what methods of bird damage control – chemical, acoustical or visual – are used by growers in the USA and what measure of success is achieved.

Rogers: Methiocarb is used fairly extensively on sprouting corn and is registered for use on sour and sweet cherries. Avitrol (4-amino-pyridine), with which some of you may be familiar, is used as a bait in which one in a 100 particles is treated with the toxin. The bait is applied in swathes from the air to fields of sunflower and sweetcorn. It does kill birds but before they die, they go through a mad flight, emitting distress calls and this causes the rest of the flock to leave the fields. In addition to these chemical means there is the usual run of netting, carbide and propane exploders and then farmers own techniques including scarecrows, tin cans or what-have-you.

Higgons: The applied nature of the "art" is very similar to that in Europe and there is room for much improvement. Can I ask if anything is used deliberately to cause bird deaths? This would not appear to be allowed under the EEC and UK regulations; is the use of poisons legal in the USA?

Rogers: The use of avitrol is allowed in the USA because it kills only a small proportion of the flock and the behaviour of affected birds reinforces the playing of alarm and distress calls. Also I forgot to mention the use of strychnine baits and Rid-a-Bird perches; these contain a contact poison, which is absorbed through the bird's feet, and they are registered for indoor use in warehouses against sparrows and pigeons.

Bunyan: Could Dr Rogers tell us something about levels of methiocarb required to produce conditioned aversion; is there a no effect level and what sort of residues are found on treated crops?

Rogers: I am certain there is a no effect level but I have never done a dose/response kind of experiment. In the laboratory I have used 0·07% technical methiocarb but in the field a 75% wettable powder is used at a rate of 2 lb a.i./acre. I believe the residue tolerance level is set at 0·1 ppm with a minimum interval of seven days between last application and harvest. A maximum of three applications per season is allowed. Many of the treated cherries go for processing so there is not that much of a problem concerning the acceptability of treated cherries to the consumer but I have heard that some growers are increasing the pre-harvest interval because of the cosmetic aspects of residues.

Drent: Dr Green showed us that damage to the sugarbeet crop is greatest at intermediate weed densities and he suggested a bell-shaped damage function in relation to seed density. My question is, can you work out the loss of yield a farmer can accept, from allowing weeds in his field which, as I understand, would lessen damage by skylarks and thus have the same effects as treating with aldicarb?

Green: At the moment it would be uneconomic to use aldicarb to prevent skylark damage because we cannot predict where damage will occur and prophylactic treatment of all fields would be very expensive. I do not know whether one would get an economic return by modifying herbicide treatments; we would need a lot more data about weed competition with the crop before we could work that out.

Haskell: What would happen if you saturated the ground with artificial weed seeds?

Green: My work suggests you would get a lot of birds in the fields but little damage to beet. Perhaps another way of achieving much the same effect would be to use a herbicide that prevented seeds from germinating, thus preventing the natural reduction in seed density. On another point; I have some indication that the palatability of aldicarb-treated seedlings varies with age but as yet I have no hard data. In cage tests skylarks would eat the germinating seedlings as they broke the soil surface but not when they were fully emerged, and I wonder if this has something to do with the rate of uptake of aldicarb, which must be dependent on the development of the root system and therefore takes a little time.

Bunyan: That seems fair comment. In our experimental work aldicarb residues were found to build up to a maximum but we also found aldicarb to be extremely mobile in the soil, and weeds, both in and out of the row, contained residues similar to those in sugarbeet seedlings. We also detected aldicarb residues in skylarks up to 60 days after seedling emergence.

Stanley: Information presented at Zurich last summer indicated that the mobility of aldicarb varies greatly with soil type and moisture level so whether you get residues restricted to the row would depend very much on conditions.

Dunning: Aldicarb is used commercially on about 33% on the English beet crop (200 000 ha) and other carbamates on a further 17%. Thus 50% is treated, principally against nematodes, soil insects and seedling foliage pests. The decrease of skylark grazing is thus a bonus but it is not clear whether deliberate application to obtain this effect would be an "approved" treatment. Beet are grown on 20 in. rows; the aldicarb is placed in the seed furrow (0·5–1 kg a.i./ha) and herbicides applied to the soil surface in a 7 in. band over the row. Weed seedlings grow between the rows and are later removed by tractor hoeing; we are trying to discourage the commercial trend to treat the whole soil area with herbicides.

Muir: Our work at East Malling Research Station on chemical repellents is done with captive bullfinches and we have found that primary taste materials were not particularly repellent, but substances causing skin irritations were much more effective. Anthraquinnone, dithiocarbamates, organo-tin compounds and a number of fungicides that have been reported to cause skin sensitisation, all show some evidence of a repellent effect and it is an area where more research could usefully be done.

Cuthbert: With reference to Mr Wright's remark about the archaic use of manure mixtures as repellents, I should like to draw attention to the modern practice of applying slurry to pastures which, it has been suggested, appears to deter grazing by geese. These observations are purely subjective but I would be interested to know if anyone else has views on this practice.

Brough: I am not able to comment on the repellency of slurry to geese but I do know that the application of slurry attracts gulls which are believed to feed on invertebrates flooded out of the soil rather than any components of the slurry itself.

Milne: I would agree with John Cuthbert that slurry does seem to deter geese.

Drent: In Holland we have a high rainfall and applications of slurry may deter barnacle and brent geese for three or four days but certainly not more.

Stanley: So often chemical repellents are evaluated in trials involving a choice, whereas under normal conditions, when whole fields are treated, no choice will exist. Can this situation be improved?

Rogers: I think the answer is unquestionably "yes". We have for a long time thought the best design was to treat whole fields, with other fields serving as controls, but with all the difficulty of predicting where damage will occur the sample size tends to be small. We now favour a split plot design – treating half of each field and leaving half as control.

Evans: I feel what needs to be stressed, when considering field tests, is the condition of the bird itself. A repellent may prove effective against a species at a particular time or place yet fail against the same species at another time or in a different situation because, for example, the bird was building up its protein reserves. Such factors need to be borne in mind in field trials.

Wright: That describes exactly what is found throughout the literature on repellents – an inability to repeat experiments, in time and space, with comparable results.

Higgons: In experiments carried out by my Company about three years ago on fruit and brassica crops we found some superb repellents when we were working on plot sizes of 10 or 50 m^2 but as we increased the plot size to above an acre, so the repellency declined, and at two or three acres it had disappeared. I am sure it is a reaction to the degree of distaste, and the inconsistency of results is related to plot size and plot design.

Smith: Can anyone report any progress in identifying the active compound in insects that are repellent to birds? I know Dr Miriam Rothschild was working in this field at one stage.

Wright: Some compounds have been identified, the cardiac glycosides, for instance, but although I feel sure work is still going on I am not aware of any recent publications.

Haskell: I think this is a very interesting point indeed, there are a host of things you can think of against this physiological background. One of the things we are looking at in insect control is the question of finding chemicals that block the sensory receptors; we know that such chemicals exist and I should have thought, if enough work was done, it would be possible to find chemicals that would block the sense of taste in birds. We are finding there are very few insects that actually have receptors tuned to particular chemicals; most insects take in a lot of information and process it in the CNS, locusts for instance have something like 40–50 000 neurons used for sensory receptors connected with feeding. I think birds must have an even more elaborate system.

Bransden: Several years ago Rachel Carson wrote a book called *Silent Spring* condemning American farmers for everything they had ever done. Yesterday we were talking about the EEC Directive on birds and again we farmers are blamed for the lack of plants, butterflies, birds, in fact everything. But they are now doing in America things that are forbidden in England (e.g., use of avitrol) and I should like to ask Dr Rogers if public opinion in America has come full circle.

Haskell: I thought Rachel Carson had been deified in the USA in the form of the EPA.

Rogers: No, public opinion has not really changed. As Dr Haskell just intimated, Rachel Carson resulted in the EPA and the major result of her work, I think, has been the entire environmental movement in our country, the creation of the EPA and the regulatory mechanism that EPA enforces, so that whatever is used must pass more or less rigorous examination. Though there may be hazards, they are known and balanced against the benefits.

Wright: I, too, should like to comment on Mrs Bransden's point because, in many respects, she has put her finger on the pulse of the situation and drawn attention to the fact that birds are a political issue in this country. It is important to realise that decisions on what techniques may or may not be used against birds are taken on political rather than scientific grounds and if farmers feel they are unnecessarily restricted in what they can do to control birds then they should make their views known through the National Farmers' Union which is able to represent them at a political level.

Higgons: Dr Rogers mentioned assessment of risk (of a pesticide) against benefit. In our dealings with the EPA recently one feature we noticed which is totally different to the UK, and a very good thing, is the scientific and logical assessment of risk against benefit. If the benefits can be shown to override the risks, then there is a more logical movement towards registration of the product even if it may necessitate the sacrifice of one or two holy cows. I think this risk-benefit analysis, based on logical and public debate, is something that has not yet entered into British politics.

Haskell: Well, it seems to me it is time to try to bring this Symposium to an end. As a member of the British Crop Protection Council, I can say that in sponsoring this Symposium we recognised that there were special problems about bird control, not least of which are the political aspects just raised; from the point of view of the Centre for Overseas Pest Research, I can say that we have tremendous difficulties of that sort in overseas countries too. It seems to me that when we talk about preventing damage, which is the real thing we are after, killing birds is out, because it is non-selective, probably ineffective, illegal in many cases and emotionally and politically bad, anyway. We have heard in this Symposium about potential damage control measures using acoustic and visual scarers, chemical repellents, alternative food and so on and it may be felt that BCPC has not included much about bird control. But the whole point was to expose the problems surrounding bird control and find out what lines of research should be followed in the future in order to develop new techniques that would be emotionally, politically and legally acceptable. I hope the publication of the proceedings of the Symposium will serve to stimulate research on these lines and I feel that research on both the physiology of birds and behavioural ecology are two of the most important aspects to follow up. Finally, I ask you to join me in thanking the speakers this afternoon for a very interesting Session indeed.

Index

Contributors

Part 1 General Considerations

Chairman **E. N. Wright,** Ministry of Agriculture, Fisheries & Food, Worplesdon Laboratory, Guildford, Surrey GU13 3LQ, England.

Speakers **N. W. Moore,** Nature Conservancy Council, 19/20 Belgrave Square, London SW1X 8PY, England.

J. J. M. Flegg, East Malling Research Station, East Malling, Maidstone, Kent ME19 6BJ, England.

H. Boyd, Canadian Wildlife Service, Ottawa, Ontario, Canada K1A 0E7.

A. W. Colling, Biel Mill Cottage, by Dunbar, E. Lothian, Scotland.

Part 2 Starlings as Agricultural Pests

Chairman **G. M. Dunnet,** Department of Zoology, The University, Aberdeen, Scotland.

Speakers **C. J. Feare,** Ministry of Agriculture, Fisheries & Food, Worplesdon Laboratory, Guildford, Surrey GU3 3LQ, England.

J. Tahon, Ministere de l'Agriculture, Station de Zoologie Appliquee de l'Etat, Centre de Recherche Agronomique, Chemin de Liroux, 8, B 5800 Gembloux, Belgium.

R. F. Verheyen, Department Biologie, Universitaire Instelling Antwerpen, Universiteitsplein 1, B 2610 Wilrijk, Belgium.

J. M. Tinbergen and R. H. Drent, Zoological Laboratory, University of Groningen, Haren, Netherlands.

Part 3 Bird Scaring

Chairman **J. R. Krebs,** Edward Grey Institute of Field Ornithology, The University, South Parks Road, Oxford, England.

Speakers **P. J. B. Slater,** School of Biology, University of Sussex, Brighton, England.

J-C Brémond, Laboratoire d'Ethologie Experimentale, Hameau de Saugis, 28210 Nogent-le-Roi, France.

I. R. Inglis, Ministry of Agriculture, Fisheries & Food, Worplesdon Laboratory, Guildford, Surrey GU3 3LQ, England.

M. Owen, The Wildfowl Trust, Slimbridge, Gloucester GL2 7BT, England.

Part 4 Chemical Bird Repellents

Chairman **P. T. Haskell,** Overseas Development Administration, Centre for Overseas Pest Research, College House, Wright's Lane, London W8 5SJ, England.

Speakers **E. N. Wright,** Ministry of Agriculture, Fisheries & Food, Worplesdon Laboratory, Guildford, Surrey GU3 3LQ, England.

J. G. Rogers, Jr., U.S. Fish and Wildlife Service, Division of Wildlife Research, Washington D.C. 20240, U.S.A.

R. Green*, Brooms Barn Experimental Station, Higham, Bury St. Edmunds, Suffolk IP28 6NP, England.

**Current address: The Game Conservancy, Fordingbridge, Hampshire, England.*

L. E. Fellows, Jodrell Laboratory, Royal Botanic Gardens, Kew, Richmond, Surrey TW9 3A13, England.

Delegates

D. A. W. Alexander
ADAS Land Service, MAFF, Great Westminster House, Horseferry Road, London.
F. Al-Jaborae
Edward Grey Institute, Zoology Department, Oxford University, Oxford.
C. E. Armstrong
Ash Lodge, Fulbeck, Grantham, Lincs.
S. Bransden
B. E. Bransden & Sons Ltd., Lyne Farm, Chertsey, Surrey.
D. M. Broom
Department of Zoology, University of Reading, Reading, Berks.
T. Brough
MAFF, Worplesdon Laboratory, Guildford, Surrey.
P. J. Bunyan
MAFF, Slough Laboratory, London Road, Slough, Berks.
C. J. Cadbury
RSPB, The Lodge, Sandy, Bedfordshire.
K. Charman
Nature Conservancy Council, Calthorpe House, Calthorpe Street, Banbury, Oxon.
D. C. M. Corbett
Agricultural Research Council, 160 Great Portland Street, London.
R. J. Cowie
Department of Zoology, South Parks Road, Oxford.
J. H. Cuthbert
DAFS, Agricultural Scientific Service, East Craigs, Edinburgh, Scotland.
R. A. Dunning
Broom's Barn Experimental Station, Higham, Bury St. Edmunds.
P. J. Edwards
ICI Ltd., Jealott's Hill Research Station, Bracknell, Berks.
P. G. H. Evans
Edward Grey Institute, Zoology Department, South Parks Road, Oxford.
R. Forrester
University of Reading, Zoology Department, Reading, Berks.
H. J. Gould
ADAS, Block III, Burghill Road, Westbury on Trym, Bristol.
P. Gramet
INRA, CNRZ, Jouy-en-Josas, France.
J. Harradine
WAGBI, Marford Mill, Rossett Clwyd, North Wales.
D. G. Higgons
The Boots Company Ltd., Agricultural Research, Pennyfoot Street, Nottingham.
J. M. Hoyoux
Station de Zoologie Appliquee, Chemin de Liroux 8, 5800 Gembloux, Belgium.

T. Johnston
The Edinburgh School of Agriculture, West Mains Road, Edinburgh, Scotland.
P. J. Jones
COPR, College House, Wrights Lane, London.
A. K. M. St. Joseph
Wildfowl Trust, Slimbridge, Gloucs.
D. A. Kendall
Agricultural Research Council, Long Ashton Research Station, Long Ashton, Bristol.
I. F. Keymer
MAFF, Veterinary Investigation Centre, Jupiter Road, Norwich, Norfolk.
J. Lazarus
Department of Psychology, The University, Newcastle upon Tyne.
R. W. Lemon
Fisons Ltd., Chesterford Park Research Station, Saffron Walden, Essex.
E. Lester
Rothamsted Experimental Station, Harpenden, Herts.
P. J. Long
Fieldspray Ltd., Feering, Colchester.
G. M. Mackintosh
Division of Agricultural Zoology, School of Agriculture, 581 King Street, Aberdeen, Scotland.
J. MacLennon
Culterty Field Station, Newburgh, Aberdeenshire, Scotland.
A. D. Martin
MAFF, Tolworth Laboratory, Surbiton, Surrey.
N. J. Mathews
Zoology Department, East Malling Research Station, Maidstone, Kent.
G. V. T. Matthews
Wildfowl Trust, Slimbridge, Gloucestershire.
H. J. Midwinter
Little Massatts, Milton Hill, Abingdon, Oxon.
H. Milne
Culterty Field Station, Newburgh, Aberdeenshire, Scotland.
F. J. Moore
MAFF, Great Westminster House, Horseferry Road, London.
J. W. S. Mount
S. W. Mount & Sons Ltd., Little Barton Farm, Canterbury, Kent.
R. C. Muir
East Malling Research Station, Maidstone, Kent.
J. W. Newbold
Zoology Department, West of Scotland Agricultural College, Auchincruive, Ayr, Scotland.
R. J. O'Connor
British Trust for Ornithology, Beech Grove, Tring, Hertfordshire.

P. Osborne
East of Scotland College of Agriculture, West Mains Road, Edinburgh, Scotland.
G. G. Pope
COPR, Porton Down, Salisbury, Wiltshire.
B. D. Smith
Agricultural Research Council, Long Ashton Research Station, Bristol.
P. I. Stanley
MAFF, Tolworth Laboratory, Surbiton, Surrey.
K. Swannack
MAFF, Bridgets Experimental Husbandry Farm, Martyr Worthy, Winchester.
R. J. P. Thearle
MAFF, Tangley Place, Worplesdon Laboratory, Guildford, Surrey GU3 3LQ.
B. H. Theobald
Westbury Farm, Purley, Reading, Berks.
H. V. Thompson
MAFF, Worplesdon Laboratory, Guildford, Surrey GU3 3LQ.
O. G. Williams
Advisory Committee, Protection of Birds, Meinl Gwynidn, Brynsiengyn, Anglesey.
E. K. Woodford
Bear Farm, Binfield, Bracknell, Berks.

The British Crop Protection Council

The British Crop Protection Council (BCPC) exists to promote the knowledge and understanding of crop protection. It was founded in 1968 when the British Weed Control Council, set up in 1953, and the British Insecticide and Fungicide Council, set up in 1962, merged to form a single body concerned with all aspects of crop protection. The BCPC is essentially a British organisation but its work is rapidly becoming international in outlook.

The Council is composed of corporate members including Government bodies, research and advisory services, the farming and agrochemical industries, distribution and contracting services, industries, distribution and contracting services, environmental bodies and other organisations, as well as individual members with special qualifications and experience in the field of crop protection. This blend is probably unique.

BCPC Objectives

Members of the BCPC have a common objective – to promote and encourage the science and practice of pest, disease and weed control, and allied subjects both in the UK and overseas. To achieve this, the Council aims:

to compile and arrange the publication of information and recommendations on crop protection for specialists;

to help the public to understand the nature of pests, diseases and weeds, and their control, and the part their control plays in food production;

to provide a forum for discussion at conferences and other meetings on matters relating to crop protection and to publish and distribute the proceedings of these meetings;

to identify short- and long-term requirements for research and development in the field of crop protection;

to act as a liaison agency and to collaborate with other organisations with similar objectives.

BCPC Members

The corporate members are

Association of Applied Biologists
41 Queen's Gate, London SW7 5HU

Agricultural Research Council
160 Great Portland Street, London
W1N 6DT

British Agrochemicals Association
Alembic House, 93 Embankment,
London SE1 7TU

National Association of Agricultural Contractors
Huts Corner, Tilford Road, Hindhead,
Surrey GU26 6SF

National Farmer's Union
Agriculture House, 25/31 Knightsbridge,
London SW1X 7NJ

Natural Environment Research Council
Alhambra House, 27/33 Charing Cross
Road, London WC2 0AX

Society of Chemical Industry
Pesticide Group
14 Belgrave Square, London SW1X 8PS

UK Agricultural Supply Trade
Association Ltd.
3 Whitehall Court, London SW1A 2EQ

Department of Agriculture & Fisheries for Scotland
St Andrew's House, Edinburgh EH1 3D4

Department of Agriculture, Northern Ireland
Dundonald House, Upper Newtownards
Road, Belfast BT4 3SB, Northern Ireland

Department of the Environment (Great Britain)
2 Marsham Street, London SW1 3EB

Ministry of Agriculture, Fisheries and Food

Pesticides Branch
Great Westminster House, Horseferry
Road, London SW1P 2AE

Agricultural Development and
Advisory Service (Headquarters)
Great Westminster House, Horseferry
Road, London SW1P 2AE

Agricultural Development and
Advisory Service
The Harpenden Laboratory, Hatching
Green, Harpenden, Herts AL5 2BD

Ministry of Overseas Development
Eland House, Stag Place, London
SW1E 5DH

Further information about the BCPC, its organisation and its work can be obtained from

The Administrative Secretary
The British Crop Protection Council
144–150 London Road, Croydon CR0 2TD

The BCPC publications are available from

'Shirley', Westfield Cradley, Malvern,
Worcestershire WR13 5LP